Planetary Health - Laws, Policies and Science on the 'One Health' Approach

Giovanni Antonelli • Fabrizio Penna •
Eeshan Chaturvedi • Antonio Cilento
Editors

Planetary Health - Laws, Policies and Science on the 'One Health' Approach

Springer

Editors
Giovanni Antonelli
Ministry of Culture
Rome, Italy

Eeshan Chaturvedi
Emmett Interdisciplinary Program in Environment and Resources
Stanford University
Stanford, CA, USA

Fabrizio Penna
Italian Ministry of Environment and Energy Security
Rome, Italy

Antonio Cilento
Università Mercatorum
Rome, Italy

ISBN 978-3-031-90620-6 ISBN 978-3-031-90621-3 (eBook)
https://doi.org/10.1007/978-3-031-90621-3

This Springer imprint is published by the registered company Springer Nature Switzerland AG
The registered company address is: Gewerbestrasse 11, 6330 Cham, Switzerland

To Emanuele, Carlo, Nuccio and Fabio
figli d'Italia e amici che mi mancavano
alla casualità foriera di cose belle
Giovanni Antonelli

Preface

The holistic One Health vision, i.e. a health model based on the integration of different disciplines, is both ancient and current. It is based on the recognition that human health, animal health and the health of the ecosystem are inextricably linked.

It is officially recognised by several governments, the European Commission and all international organisations as a relevant strategy in all fields that benefit from collaboration between different disciplines (law, medicine, economics, sociology, etc.).

One Health is an ideal approach to achieving environmental goals and global health because it addresses the needs of the most vulnerable populations on the basis of the intimate relationship between their health, the health of their animals and the environment in which they live, considering the broad spectrum of determinants that emerge from this relationship.

Since the covid-19 pandemic, legislation and policy have increasingly referred to One Health as a tool of transition towards sustainable systems.

To this end I assume necessary bridging the distance between law and natural science.

After my volumes '*Environmental Law Before the Courts: A US-EU Narrative*' and '*Biodiversity Laws, Policies and Science in Europe, the United States and China*', Planetary Health—Laws, Policies and Science on the 'One Health' Approach aims at offering some contributions on One Health.

Rome, Italy Giovanni Antonelli

Preface

Contents

One Health and Constitution: The Italian Case 1
Alessandro Lancione and Gianluca Di Egidio

The Sustainable Use of Healthy Soil as a Cornerstone of the One Health Approach: An Analysis of European Public Strategies and Market-Based Environmental Protection Tools 19
Alessia Depietri

Climate Change and a Comprehensive Reparation of Damaged Interests: A Cultural Issue 43
Anna Malomo

Environment, Climate and Criticality of the Alleged Category of So-Called "Common" Goods 53
Giovanni Perlingieri

Reverse Correlation Between Allergic Pollens and Fungi Amounts Annually with Allergic Sensitization Rate to Them 65
Young-Jin Choi, Kyung-Suk Lee, and Jae-Won Oh

California's Wildfire Crisis and the Future of Planetary Resilience 81
Jessica Yu and Eeshan Chaturvedi

Fundamental Human Rights and the Climate Crisis: The Critical Interaction Between Law and Medicine 111
Andrea Rodgers, Anpotowin Jensen, Vivian Vo, and Kajal Khanna

One Health: Oriented Directions to Bring Science to Policy in Climate Change Mitigation and Adaptation 125
Skye S. Flanigan, Sweta Waghela, Vanitha Sampath, and Kari C. Nadeau

Indigenous Knowledges, One Health, and Planetary Health: Building Bridges for the Response to Climate Change 151
Brittany Bingham, Nicole Redvers, Sean Hillier, and Tamara Riley

One Health and Constitution: The Italian Case

Alessandro Lancione and Gianluca Di Egidio

Abstract The *One Health* approach, which calls for integrated protection of human, animal and environmental health, necessarily requires intervention at a regulatory level.

In particular, the role of legislation is indispensable in order to allow effective implementation of the approach: as was declared by the FAO in 2020 "legislation is a powerful means by which countries and regional organizations translate the One Health objectives into concrete, sustainable and enforceable rights, obligations and responsibilities".

Starting from these considerations, the paper, after having outlined the main interventions at an international and European level, intends to focus on the Italian case, where it analyzes the multiple problematic profiles connected to the implementation of this new healthcare model.

1 The *One Health* Approach and Its Legal Implications

It is certainly not easy to give an univocal definition of the *One health* approach. In scientific language, in fact, the expression *One Health* is used with multiple meanings and it is even superimposed on different health paradigms, such as *Eco Health* and *Planetary Health*, which start from different theoretical concepts.[1]

Alessandro Lancione drafted Sects. 1, 4, 4.1, 4.2, and 4.3; Gianluca Di Egidio drafted Sects. 2, 3, and 5. Both Authors drafted the conclusions (Sect. 6).

[1] 1 See Lerner and Berg (2017).

A. Lancione (✉) · G. Di Egidio
University of Teramo, Teramo, Italy
e-mail: alancione@unite.it; gdiegidio@unite.it

G. Antonelli et al. (eds.), *Planetary Health - Laws, Policies and Science on the 'One Health' Approach*, https://doi.org/10.1007/978-3-031-90621-3_1

In general terms, this approach represents a health model based on the awareness of the close link existing between humans and all other animal species.[2]

This connection between human being and animals was highlighted by veterinary science in the nineteenth century throughout the discovery of zoonoses (i.e. infections and diseases that can be transmitted from animals to humans) and was then taken up again by Carl Schwabe, who used the expression *One Medicine* to indicate an unitary vision of human and veterinary medical sciences.[3]

Subsequently, with the advent of the twenty-first century, the diffusion of new global epidemics (such as Ebola, avian flu, Sars epidemic, swine flu), has made it clear that health protection cannot ignore the risks deriving from the "human-animal interface".[4] As a consequence, the scientific community became aware that to guarantee health at a global level it is necessary to promote new forms of protection, according to a holistic approach, which refers not only to the individuals, but also to all living parts of the planet.

This intuition dates back to the veterinarian William B. Karesh, who summarized this new health paradigm in the expression *One Health.*[5] The idea of "One health" was then taken up again in a symposium held at Rockefeller University of New York, at the end of which were formulated the so-called "Manhattan principles": a strategic document, containing 12 recommendations aimed at ensuring the prevention of epidemics and the maintenance of the ecosystems integrity, in the interests of people, animals and biodiversity. These recommendations (addressed to national and international political actors, the scientific community and civil society) are based on the recognition of an "*essential link between human, domestic animal and wildlife health and the threat disease poses to people, their food supplies and economies, and the biodiversity essential to maintaining the healthy environments and functioning ecosystems we all require*".[6]

Initially, the *One Health* approach was configured by the scientific community as an innovative strategy to face the new challenges of the future, such as food security and sustainability,[7] the fight against climate change.[8]

Subsequently, following the Covid 19 pandemic (virus of animal origin transmitted to humans), the idea of holistic approach to prevention and health protection has definitively become part of the political guidelines adopted by the main International Organizations.

---

[2] Gibbs (2014), p. 86.

[3] Schwabe (1984). For further historical insights see Zinsstag et al. (2011).

[4] Connolly (2017), p. 483.

[5] Karesh and Cook (2005), p. 50: "*Global health will not be achieved without a shift from the expert-controlled, top-down paradigm that still dominates both science and medicine. A broader, more democratic approach is needed, one based on the understanding that there is only one word – and only one health*".

[6] The complete list of Manhattan Princes can be viewed at the following link: https://oneworldonehealth.wcs.org/About-Us/Mission/The-Manhattan-Principles.aspx.

[7] Garcia et al. (2020).

[8] Black and Butler (2014).

In particular, in 2021 the World Health Organization (WHO), the Food and Agriculture Organization of the United Nations (FAO), the World Organization for Animal Health (OIE) and the United Nations Environment Programme (UNEP)[9] issued a joint declaration, from which the main aspects of the *One Health* model have emerged.[10]

First, the declaration defines the *One Health* paradigm as an "*integrated, unifying approach that aims to sustainably balance and optimize the health of people, animals and ecosystems. It recognizes the health of humans, domestic and wild animals, plants, and the wider environment (including ecosystems) are closely linked and inter-dependent*".

Secondly, the statement points out how this approach "*mobilizes multiple sectors, disciplines and communities at various level*": scientific research, investments, public policies and administrative procedures. In this sense, the *One Health* model is configured as a wide and multidisciplinary approach, which involves a plurality of subjects: not only scientists (doctors and veterinarians), but also policy makers (national and supranational).

Finally, the declaration highlights the possibility of applying the *One Health* approach at community, subnational, national, regional, and global levels, through governance and coordination mechanisms.

In particular, the last aspect also emerges from the report formulated in 2021 by the Pan-European Commission on Health and Sustainable Development, on the recommendation of the WHO Regional Office for Europe.[11] The report urges the pursuit of public policies inspired by the aforementioned approach at all levels of government. In fact, as it has been highlighted, the idea of "one health" that concerns all living beings in "one world" can only be guaranteed through intersectoral governance mechanisms, operating at international, state and sub-state levels.[12]

Starting from these considerations, the chapter focuses on the ways in which the *One Health* approach has been pursued at various levels of government. First, it explains how this approach has been implemented and promoted by international and European institutions (Sects. 2 and 3). Subsequently, it will consider the Italian case (Sects. 4 and 5), where it analyzes the multiple problematic profiles connected to the implementation of this new healthcare model.

---

[9] These international organizations form the so-called "Tripartite+ Alliance", on which the international governance of the One Health approach is based. For further details see the next paragraph.

[10] *Joint Tripartite (FAO, OIE, WHO) and UNEP Statement Tripartite and UNEP support OHHLEP's definition of "One Health"*, 1 December 2021, available at the following link: https://openknowledge.fao.org/server/api/core/bitstreams/54a0f96f-066e-4a16-a186-1a4bbd9d9303/content.

[11] Pan-European Commission on Health and Sustainable Development, *Drawing light from the pandemic. A New Strategy for Health and Sustainable Development*, available by the following link: https://www.who.int/europe/publications/rethinking-policy-priorities-in-the-light-of-pandemics%2D%2Da-call-to-action.

[12] Ragone (2022), p. 813.

2 The *One Health* Approach in the Field of International Law

In international law, the interrelationship between human health, animal health and the health of the environment in which they live was already mentioned in several health strategy documents proposed by the World Health Organisation (WHO) Since the late 1980s, among which stands out the 1986 Ottawa Charter.[13]

This document, while taking into account the environment as a fundamental factor of human and animal health, considered it in an anthropocentric view, since it is recognized as valuable because the "society one lives in that creates conditions that allow the achievement of health by all its members".[14]

Later, in September 2004, that is to say, after the outbreak of SARS,[15] a group of strategic thinkers and health experts from all over the world met for a convention entitled "Building Interdisciplinary Bridges to Health in a Globalized World", organized by the Wildlife Conservation Society. In particular, this conference focused on the interconnections between diseases of humans, pets and wildlife, then proposed the "Manhattan Principles on One World, One Health".[16]

In the following years, these indications were taken up and valorized by both the American Medical Association that, in 2007, approved the resolution "One Health", promoting an initial hypothesis of partnership between human medicine and veterinary medicine, representatives from over 100 countries and 29 international organizations, gathered in New Delhi, India, for the International Ministerial Conference on avian or bird flu, caused by the H5N1 virus. The work of this group formed the basis for the introduction of the 2008 common strategic framework, adopted by the Food and Agriculture Organization of the United Nations (FAO), the World Organisation for Animal Health (WOAH) and the World Health Organization (WHO) to provide a common response, at international level, to the risks of developing new diseases, including pandemic-type.[17]

---

[13] Ottawa Charter for Health Promotion, 1986, https://www.who.int/teams/health-promotion/enhanced wellbeing/firstglobal-conference The first International Conference on Health Promotion, convened in Ottawa on 21 November 1986, presents this Charter to stimulate action for Health for All for the year 2000 and beyond. This conference was primarily a response to the growing global expectations for a new public health movement. The discussions focused on needs in industrialised countries, but looked at similar issues in all other regions. The conference built on progress made through the Declaration on Primary Health Care of The 1978 O.M.S. document on the objectives of Health for All and the recent debate on intersectoral action in favour of health held at the World Health Assembly.

[14] Ottawa Charter for Health Promotion, ib., § Moving into the future, p. 4.

[15] The SARS epidemic of 2002–2004 was caused by the spread of a respiratory disease caused by the coronavirus SARS-CoV. Probably started in November 2002 in China, the spread of the virus affected about thirty countries of the world. The identification of the coronavirus that causes SARS was made possible thanks to the research of the Italian doctor Carlo Urbani.

[16] For an overview of the contents of the Manhattan Principles, see para 1.

[17] Cf. Latino (2022), p. 790.

Through these documents, the *One Health* theory is gradually being significantly upgraded, becoming not just an approach recommended by health technicians and experts, a model adopted by the main Organizations that make up health governance at global level.

As we approach 2010, the Hanoi Declaration makes the *One Health* approach a reference paradigm for health prevention.[18] In fact, the document mentioned recommends its widest possible application, thus moving from a purely scientific and theoretical vision to the practical application of *One Health*, by identifying concrete actions. The contents of the 2010 Declaration are immediately taken up and made their own by the UN and the European Union.[19]

However, the recommendations and international policy statements have been largely ignored by Member States or only partially adopted over the years. It is difficult to abandon the idea that man is the centre of all preventive and health protection action; therefore, until a few years ago, actions for the direct protection of animal life and environmental resources were left on the margins of state policies.

A strong impetus to the concrete implementation of the *One Health* approach, as already suggested by the programmatic lines of international governance in the field of health, came from the pandemic crisis caused by Covid-19.

The expansion of the Covid-19 virus was the event that, with the utmost urgency, required to rethink existing health prevention systems without further delays than those already recorded in the recent past. Systems that, according to the already mentioned logic *One Health*, should be designed in a global way, enhancing the interconnections between individuals, animals and environment.[20]

At present, there are many international organisations involved in this field; the most important of these are: the World Health Organization (WHO), the World Organisation for Animal Health (WOAH), the Food and Agriculture Organization (FAO) and, finally, the United Nations Environment Programme (UNEP). All these international organizations are, in various ways, attributable to the UN.

Although they have some common goals, at least in the health sector, the Entities that define and manage the *One Health* approach (c.d. Tripartite+) differ in terms of their origins, structures and statutory purposes.[21] This observation, at the legal level, is reflected in the type of regulatory acts that they may adopt to achieve their institutional objectives. However, in general terms it is mainly soft law, with all that this

[18] This statement can be found at https://www.woah.org/app/uploads/2021/03/final-concept-note-hanoi.pdf.

[19] Identifies the Hanoi Declaration as a turning point in the development of the *One Health* approach, Latino (2022), p. 790. Since then, international meetings have also multiplied, starting with the first International Congress on the *One Health* approach in 2011 in Melbourne and the first African Conference on the *One Health* approach in Addis Ababa.

[20] On the design criteria and perspectives of health prevention systems see Peters, p. 85 ss.

[21] Cf. Flaminia (2022), p. 14.

entails in terms of their binding force vis-à-vis the States addressed by these provisions.

Looking at the current tools for implementing the *One Health* approach, in May 2021 the WHO adopted a resolution calling for "to develop and strengthen existing cooperation with FAO, OIE[22] and UNEP for the definition of a common One Health strategy". Following this resolution, on 17 October 2022 the four Tripartite+ organisations developed a joint action plan: "One Health (OH-JPA) 2022–2026".[23]

The plan includes six main actions to be implemented: strengthening countries' capacity to strengthen health systems under a *One Health* approach; reducing risks arising from the emergence or re-emergence of zoonotic epidemics and pandemics; control and eliminate endemic zoonoses, neglected or vector-borne tropical diseases; strengthen food safety risk assessment, management and communication; curb the silent pandemic of antimicrobial resistance; Strengthen the integration of the environment in the *One Health* approach.[24] Overall, these are concrete objectives that have a unitary matrix, represented by the protection of health in an integrated way, without separating the aspect of human health from that of animals and the health of the environment in which they both live.

This action plan is also a political document with no binding character; therefore, its effective implementation through the legislative intervention of transposition by individual States is still required.[25] This aspect was well highlighted by the FAO, which, already in 2020, had stressed the central role of the legislative activity of the States to translate the objectives of the *One Health* model into concrete rights, obligations and responsibilities.

In general, therefore, even the most recent international acts adopted in the context of *One Health* contain essentially soft law provisions, not endowed with direct binding force. However, this does not mean that these rules are irrelevant and that they cannot produce practical effects even in the absence of state intervention to implement them.[26] Indeed, precisely because of the importance of the international organizations that have adopted them and the high rate of spontaneous adherence, the principles on *One Health* seem to be able to be inserted among the sources of customary international law.[27]

---

[22] The acronym refers to the Office International des Epizooties. It is simply the former name of the World Organisation for Animal Health.

[23] This plan can be found at https://www.who.int/publications/i/item/9789240059139.

[24] On the objectives and contents of the plan "One Health (OH-JPA) 2022–2026", cf. Iacovino and Randazzo (2023), pp. 302 and 326. Important information on the plan under consideration is also available on the website of the Istituto superiore di sanità italiano: https://www.epicentro.iss.it/globale/ohjpa-2022-2026.

[25] The Action Plan also seems to be on a line of continuity with the general framework set out in the Sustainable Development Goals identified by the UN 2030 Agenda for Sustainable Development (https://www.un.org/sustainabledevelopment/).

[26] See Latino (2022), p. 801.

[27] On the subject of environmental principles in international law, see Boissons De Chazournes and Maljean Du Bois (2016), pp. 1 ff.; Fajardo, pp. 38 ff.

According to this reconstruction, within the framework of the Italian legal system, the principles relating to the *One Health* paradigm would benefit from constitutional coverage pursuant to article 10, paragraph 1, Cost.

3 One Health and EU Law

Like most of its Member States, the European Union, while officially recognising and making political use of the international recommendations on *One Health*, which date back to the 1970s,[28] has largely disregarded their content or, at best, it has only partially implemented them.

The international scenario has changed rapidly with the arrival of the Covid-19 pandemic, the European Union, starting from the principles on the protection of the environment, human and animal health already consolidated in its system, has incorporated the *One Health* approach in the latest actions to ensure a sustainable development of the common market, including by combating climate change. These objectives are currently a priority in the process of European integration.

Human health, animal welfare and a healthy environment are all areas which fall within the EU's remit in various ways.[29]

At the Paris Summit of July 1972, promoted by French President Pompidou, the Heads of State and Government of the then member countries of the European Community recognized that in the context of economic expansion and improvement of the quality of life, The environment needed to be given special attention. An environmental action plan followed, and on the basis of it a series of measures adopted thanks to an extensive interpretation of the aims of the Community, provided for in Article 2 of the EEC Treaty, and the reference to Article 100 TCEE on the approximation of the laws of the various Member States.[30]

In this context, the role of the case law of the EC Court of Justice, which had declared, already around the middle of the 1980s, that environmental protection was one of the essential aims of the Community, should also be recalled.[31]

With the entry into force of the Single European Act in 1986, the Community's environmental policy was finally incorporated into the European Treaties as primary law sources; it thus became legitimate, with a specific legal basis, The EC's action to protect the environment.[32]

---

[28] On international health strategy documents, cf. Sect. 2.

[29] On the Union's competence in these matters, see Sanna, pp. 33 ff.

[30] For an analysis of the Union's environmental policy, see Costato and Manservisi (2012); Kramer (2016).

[31] Cf. Court of Justice, C. No. 240/83, point 13. 7 February 1985.

[32] Precisely as a result of the entry into force of the Single European Act, in the subsequent sentence 20 September 1998, C. 302/1986, the Court stated that since environmental protection is an essential objective of the European Community, it constitutes a requirement which may also place a limit on the free movement of goods (Art. 30 TCEE, now Art. 34 TFUE).

In the following years, with the various revisions of the Treaties, the discipline was progressively extended, allowing a substantial production of Community legal norms, also because the environment was considered as a matter of concurrent competence (Art. 4, para. 2, TFEU). It should also be stressed that, under Article 11 TFEU, where the principle of integration is laid down, environmental protection needs must always be integrated into the definition and implementation of other Union policies, In particular, to promote sustainable development.[33]

The protection of human health is also a part of environmental protection under EU law. The Treaty of Rome already clearly stated that there is a strong link between the environment and human health, and that therefore an integrated policy for the protection of these two legal assets must be developed.[34]

With the 1992 Maastricht Treaty, human health protection takes on an autonomous relevance in primary EU law: from now on, a "high level of human health protection" must be guaranteed in all EU policies, especially as regards cross-border risks.[35]

However, it should be specified that the cases in which the European Union can adopt binding measures to protect human health are rather limited, this possibility is provided for only in relation to the quality and safety standards of organs and substances of human origin, blood and blood products, and in the veterinary field, always with the aim of protecting public health. In all other cases of risk, the EU can only support and coordinate actions by the Member States, without being able to limit their competence in health policy choices.[36]

The EU's environmental competence has also been very important in developing animal health and welfare interventions aimed at conserving biodiversity. In more detail, the EU has signed several international conventions for the conservation of natural habitats and wildlife, for the protection of wild animals from illicit trade, For the welfare of animals used in research and for the protection of the marine ecosystem.[37] The EU has also adopted important regulations and directives for the Member States in these areas.[38] Animal health is also guaranteed by European food safety

[33] The Lisbon Treaty also included among the objectives of the European Union the fight against climate change, through the Green Deal, now a reference point for the process of European integration.

[34] More specifically, the Treaty of Rome established health protection as an exception which could be used by Member States to restrict the free movement of goods within the single market (Art. 36 TCEE).

[35] Cf. Art. 129 TEC, now Art. 168 TFEU.

[36] On this subject, see. De Grove-Valdeyron (2018); Di Federico and Negri (2020).

[37] International conventions on the subject are listed by Violini, p. 36 s.

[38] Among the many EU legislation, it is worth mentioning: the Directive on the conservation of natural habitats and wild fauna and flora (dir. 92/43/EEC); the Directive on the protection and control of wild birds, including rules on sustainable hunting (dir. 2009/147/EC); the Regulation for the protection of wild fauna and flora species by controlling their trade (Reg. EU 2017/160); the Directive on the protection and conservation of the marine environment (Dir. 2008/56/EC). Finally, with the EU Regulation 2016/429, the Union has approved new provisions on animal health and welfare in order to prevent and contain the effects of animal diseases, enhancing early diagnosis.

rules: general provisions on the monitoring and treatment of infectious diseases and their vectors; to avoid possible risks of contamination of the food chain.

The evolution of European legislation has reached a turning point with the approval of the 2017 One Health action plan against antimicrobial resistance.[39] The measures adopted under this plan are not binding on Member States and citizens, as they almost always consist of guidelines and strategies; However, they are of considerable importance because they tend to orient the legislation of individual Member States towards the adoption of the One Health approach.

The evolution of European legislation has reached a turning point with the approval of the 2017 One Health action plan against antimicrobial resistance. The measures adopted under this plan are not binding on Member States and citizens, as they almost always consist of guidelines and strategies; However, they are of considerable importance because they tend to orient the legislation of individual Member States towards the adoption of the One Health approach.[40]

This approach is also enhanced in the EU Health Programme 2021–2027, established by the European Regulation 522/2021 with a significant financial allocation (about 2.5 billion euros). In this document, the One Health is defined as a model that recognizes the connection of human health with animal health and the health of the environment; Therefore, all measures to counter health threats must take into account the deep link between these three dimensions.[41]

Finally, in 2021 the European Commission decided to establish the European Health Emergency Preparedness and Response Authority, which is entrusted with the task of preventing, detecting and responding rapidly to health emergencies.[42]

Therefore, adopting a global vision, among supranational bodies and, more specifically, within the institutions of the European Union has now consolidated a strong sensitivity and attention to the One Health paradigm, considered as the best approach to effectively address risks related to human health, animal welfare and protection of natural ecosystems. However, given the programmatic or, in any case, non-binding nature of acts adopted at supranational level, it is also here that individual states must apply this approach concretely in their own health policy choices.[43]

---

[39] The plan is based on the Council conclusions of 22 June 2012 on the impact of antimicrobial resistance in the field of human and veterinary health.

[40] Guidelines on the prudent use of antibiotics in veterinary medicine, guidelines on the prudent use of antibiotics in human medicine may be recalled in particular, Pharmaceutical strategy for Europe and the strategy to promote sustainable food systems.

[41] Regulation EU 2021/522 of the European Parliament and of the Council of 24 March 2021, establishing the European Union Action Programme on Health for the period 2021–2027 (EU4Health) and repealing the previous EU Regulation 282/2014.

[42] Decision of 16 September 2021 of the European Commission on preparedness and response to health emergencies.

[43] For a wider illustration of the role of the EU Member States in implementing policy-making acts adopted by the EU institutions, see Sanna, pp. 36 ff.

4 The *One Health* Approach in the Italian Legal System: Constitutional Profiles

In the Italian legal system, the three elements covered by the *One Heath* approach (human, animal and environmental health) assume their own autonomous and distinct relevance within the Italian Constitution.

4.1 The Right to Health and a Healthy Environment

Originally, only human health was recognized at constitutional level through art. 32.1, which attributes to the Republic the duty to protect health, as a fundamental right of the individual and interest of the community, and to ensure care to the indigent. According to the prevailing opinion,[44] also shared by the Constitutional Court,[45] multiple legal situations would arise from the constitutional provision: the right to freedom (consisting of the freedom to choose one's health condition and the freedom to receive treatment), the right to social benefits (the right to treatment),[46] as well as the collective interest to the health of the society.[47]

The notion of health, which is the object of various legal claims, appears much more complex. Initially, the concept of health corresponded with the psycho-physical integrity of the individual. In other words, health was defined negatively, as "absence of disease". This orientation was then overcome thanks to the work of jurisprudence, which interpreted the notion of health extensively, so as to include not only the biological, but also the social aspects connected to it.[48] This way, individual health (the object of protection of art. 32) is understood as a state of overall well-being of the person, which comprehends the work and environmental context of the subject.

Following this interpretative approach, the Court of Cassation and the Constitutional Court have derived from art. 32 of the Constitution, read in

---

[44] Pezzini (1983), pp. 25 ff.; Luciani (1991), pp. 5 ff.; Pace (2003), p. 83.

[45] Corte Cost., sent. n. 455/1990.

[46] The right to healthcare is recognized to all subjects of the Italian legal system. This "universal" model of healthcare was implemented with the law establishing the National Health Service (law No. 883/1978), which in art. 1, co. 3 establishes that "*il servizio sanitario nazionale è costituito dal complesso delle funzioni, delle strutture, dei servizi e delle attività destinati alla promozione, al mantenimento ed al recupero della salute fisica e psichica di tutta la popolazione senza distinzione di condizione individuali o sociali e secondo modalità che assicurino l'eguaglianza dei cittadini*" [*the national health service is made up of the complex of functions, structures, services and activities aimed at promoting, maintaining and recovering the physical and mental health of the entire population without distinction of individual or social conditions and according to methods that ensure the equality of citizens*].

[47] Caravita and Morrone (2016), p. 19.

[48] For further information on the evolution of the concept of health see Morana (2002), pp. 11 ff.

conjunction with art. 9.2,[49] the right to a healthy environment.[50] This jurisprudential orientation assumes the awareness (not too dissimilar to that underlying the *One Health* approach) that the protection of human health cannot be separated from the protection of the environment, in which the individual lives, given their close interconnection: it would be a "conceptual link" rather than a legal one.[51] It appears evident that, within this first perspective, the constitutional relevance of the environment is recognized according to a "functionalized" logic: in the absence of regulatory provisions, the protection of the environment is recognized by jurisprudence, in order to guarantee more effective protection of human health and well-being. This aspect clearly emerges from a decision by the Constitutional Court,[52] which configures the environment as a constitutionally protected asset as a determining element of the quality of life: its protection is not end in itself but responds to the need of ensuring a suitable habitat for human existence. In this sense, the environment would rise to a primary and absolute constitutional value.

4.2 The "Constitutionalisation" of Environmental and Animal Protection

We will have to wait for the intervention of the constitutional legislator so that the environment can find an explicit mention within the constitutional provisions.

Firstly, in this regard it is necessary to mention the Constitutional Law No. 3/2001, which modified art. 117 of the Constitution, introducing the legislative competence of the State in matters of protection of the environment and ecosystems.

More recently, Constitutional Law No. 1/2022 amended Article 9 of the Constitution, establishing that the Republic protects the environment, biodiversity and ecosystems, also in the interest of future generations and assigned to state law the task of regulating the ways and forms of animal protection. Furthermore, the law also affected Article 41 of the Constitution, which now establishes that economic initiative cannot be carried out in conflict with health and environment and entrusts the law with the determination of programs and controls for the exercise of economic activity for environmental purposes.

Therefore, through the constitutional reform, not only the protection of the environment, but also the protection of ecosystems, biodiversity and animals has found explicit constitutional recognition, becoming part of the list of principles that the Constitution defines as "fundamental".

---

[49] The art. 9.2 assigns to the Republic the task of protecting the landscape and the historical and artistic heritage of the Nation.

[50] Cass. S.U. 6.10.1979, n. 5172; Corte cost. sent. n. 167/1987.

[51] Pajno and Pucci (2020), pp. 16–17.

[52] Corte cost. sent. n. 641/1987.

In light of this constitutional intervention, the first interpretative question concerns the content of the introduced notions. In particular, the problem arises for the notion of "environment" (but also for the notion of "ecosystems" and "biodiversity"), which represents an indeterminate legal concept.

This would be a presupposed notion,[53] the meaning of which would be derived through the so-called "criterio storico-normativo":[54] that is by referring to the legal acts adopted (or in any case transposed) by the state legal system and, also, to the jurisprudence of the Constitutional Court and the supranational Courts.[55] This way, one could recall the definition offered by constitutional jurisprudence, according to which the notion of environment would express a concept comprehensive of all natural and cultural resources, configuring itself as the habitat of human beings.[56]

Under this definition, the environment constitutes the object of a specific duty of protection that the reformulated art. 9 attributes to the Republic. This provision expresses a constitutional principle,[57] addressed to all levels of government (State, Regions, and local authorities), which must exercise their powers in compliance with environmental interests.

Instead, in regards of animal protection, the constitutional legislator has not provided for direct protection, but rather has left the discipline of the forms and methods of animal protection to the discretion of the legislator;[58] this would suggest that protection could be differentiated according to the category of animals: for example, distinguishing between domestic or companion animals and animals useful to humans for various purposes (economic, religious, scientific, etc.).

4.3 *One Health and Constitution*

It should be noted that the constitutional legislator, while establishing the task of public authorities to protect the environment and animals, has not recognized the latter as holders of rights. The protection technique employed does not correspond to the eco-centric logic followed by Andean constitutionalism.[59] As it has been highlighted, the Italian Constitution, also following the aforementioned constitutional reform, places the individual (and his rights) at the center of the protection system.[60]

This, however, should not lead to the belief that the *One Health* approach, as it is inspired by a different eco-centric logic, may be incompatible with the

[53] Di Salvatore (2023), pp. 13 ff.

[54] D'Atena (1974), p. 117 ss.; Mangiameli (1986), pp. 72 ff.

[55] For the use of the "criterio storico-normativo" see Di Salvatore (2022), pp. 6 ff.

[56] Corte cost., sent. n. 12/2009.

[57] Bifulco (2022), p. 5.

[58] Vipiana (2022), p. 1117.

[59] For further information on Andean constitutionalism see Amirante (2022), pp. 111 ff.

[60] Di Salvatore (2024), p. 104.

constitutional order. In fact, it should be noted that the eco-centric conception of the *One Health* approach pertains to its ethical-value dimension and does not assume any legal relevance. In this sense, there would be no obstacle from a constitutional point of view that would prevent the implementation of legislative policies inspired by a holistic conception of health. Indeed, it has been observed that the *One Health* approach can constitute a means through which to guarantee a more effective protection of social rights and, above all, of the individual's right to health.[61]

The real problem that arises is understanding how this approach can be pursued at a regulatory level. In fact, the *One Health* approach involves matters differently placed in the division of competences between the State and the Regions, outlined by art. 117 of the Constitution: health protection is part of the list of matters attributed to the share competence between the State and the Regions; environmental protection, introduced with the aforementioned reform of 2001, belongs to the exclusive competence of the State; animal protection, on the other hand, although not explicitly placed in the division of competences outlined by art. 117, is reserved to the law of the State in accordance with the amended art. 9.

Secondly, it cannot be overlooked that the *One Health* approach can be pursued in multiple sectors (such as food security, antimicrobial resistance, mitigation and adaptation to climate change) that involve a plurality of political actors.[62]

In light of these issues, it is clear that a full regulatory implementation of the *One Health* paradigm escapes the logic of the distribution of competences by subject matter, requiring instead an approach by policy according to the decision-making model accepted in the European Union legal system.[63]

5 The Implementation of the One Health Principle in Italy. Programmatic Plans and Choices of the Legislator

In Italy, as well as in other national realities, the *One Health* paradigm has begun to arouse interest from institutional actors and to be valued by the legislator in the midst of the pandemic emergency caused by Covid 19. In fact, the reference to this approach of health policy begins to appear in the plans and draft laws presented to the Italian Parliament in 2020.

The first example is the draft law S-1891, which deals with provisions for the enhancement and strengthening of the system of prevention, planning and control in public veterinary health, with particular reference to food safety;[64] this text pro-

[61] Ragone (2022), p. 815; Galicz (2022), p. 757.

[62] On these issues see Ragone (2022), pp. 817–818.

[63] For further information on the policy approach pursued at European level see Di Salvatore (2015), pp. 405 ff.

[64] This draft bill was presented to the Senate of the Italian Republic on 22 July 2020, but it has never been examined.

posed to give concrete implementation to the *One Health* principle, adopting public health policy measures inspired by this approach, considering it "fundamental precondition for any discussion of a political nature on the subject".[65]

During this period, precisely on 6 August 2020, the National Prevention Plan (NPP) 2020–2025 was adopted with agreement between State and Regions, indicating in *One Health* the fundamental reference for addressing risks, potential or already existing, which arise from the necessary relationship between people, animals and the environment.[66] According to this perspective, the NPP 2020–2025 is also a tool for achieving the goals of the UN 2030 Agenda for the sustainable development of the planet, which requires a combined approach to economic aspects, social and environmental impacts on the well-being of people and the development of societies.

The National Plan clearly illustrates the potential of *One Health*. It is a means of addressing the cross-cutting issue of biodiversity and human health, as the document states, as well as the effective fight against antibiotic resistance—now a global problem—or the fight against the emergence of epidemics and pandemics that originate in tampering with and degradation of ecosystems, resulting in transfers of pathogens (spillover) from wildlife to domestic animals, up to subsequent transmission to humans.

The need to consider health and environment linkages in primary prevention is also highlighted by the subsequent National Recovery and Resilience Plan. In detail, mission 6 of the Plan, dedicated to health, provides for the implementation of an organic reform, so as to define a new health strategy that will allow Italy to achieve adequate quality standards of care, in line with the best European countries.[67] This new health strategy advocated in the document will increasingly have to consider the national health service as part of a wider European welfare system.

The reform planned by National Recovery and Resilience Plan, Mission 6, has been developed through two lines of action. First, the Ministry of Health, through the adoption of a regulation, has defined new organizational and technological models for the development of territorial assistance in a uniform way throughout the national territory.[68] In addition, the Italian government has adopted a specific decree-law that, among other things, has developed a new institutional framework for prevention in the health, environmental and climate sectors, in line with an integrated approach *One Health*.[69] In June 2022, this decree-law was converted into law by the Parliament.[70]

---

[65] See explanatory report Bill number S-1891.

[66] The plan can be consulted at https://www.governo.it/sites/governo.it/files/PNRR.pdf.

[67] Reform 1: Proximity networks, facilities and telemedicine for territorial health care and national network of health, environment and climate.

[68] Decree of the Ministry of Health No. 77/2022.

[69] This is the decree-law No. 36/2022, often also referred to as decree PNRR 2. The acronym PNRR is the abbreviation of National Recovery and Resilience Plan.

[70] Law No. 79/2022.

Among the contents of this law stands out the establishment of the National Health Prevention System from environmental and climate risks, configured precisely, as the new institutional and organizational framework to support the application of the *One Health* method to the policies and strategies implemented by the Italian health service with the aim of preventing, controlling and treating acute and chronic diseases, transmissible and non-transmissible, associated with environmental and climatic risks.

Pursuant to art. 27, paragraph 2, of the c.d. decree-law PNRR2, converted into law No. 79/2022, the National Health Prevention System contributes to the achievement of primary prevention objectives: health promotion, prevention and control of health risks associated with environmental and climatic factors; all this by interacting with the National System for the Protection of the Environment and applying the integrated approach *One Health*, also in its evolution *Planetary Health*. While the concept of *One Health* is more focused on the direct relationship between man, animals and environment, the latter—closer to the idea of sustainable development adopted in the 2030 Agenda—also includes in its sphere of interest social-the economic aspects of the national systems in which that particular relationship is established and developed.[71]

As regards the composition of the National Health Prevention System from environmental and climatic risks, it includes the Departments of Prevention of the Local Health Department, the Regions and the Autonomous Provinces of Trento and Bolzano, the Experimental Zooprophylactic Institutes, the Higher Institute of Health and the Ministry of Health (Art. 27, paragraph 4).

The wide range of responsibilities and the different geographical scope of the many actors involved justifies the breadth of the tasks assigned to the National Health Prevention System. The tasks of the System include the following: (a) identify and evaluate health problems associated with environmental and climate risks, to contribute to the definition and implementation of prevention policies; (b) Promote the inclusion of health in decision-making processes involving other sectors; (c) Contribute to the definition and approval of prevention programming acts and essential levels of healthcare, ensuring consistency with actions on Key Environmental Performance Levels; (d) contributing to the development of integrated criteria, methods and monitoring systems; (e) to provide support to the competent authorities for the issue of administrative authorisations or for the conduct of procedures for environmental and health impact assessment (art. 27, paragraph 3) of the Act.[72]

[71] On the differences between these two concepts, see Ragone (2022), p. 820.

[72] The health impact assessment procedure is particularly important, whereby entities whose institutional purpose is health protection are called upon to assess whether the start-up or expansion of activities of major dimensions (for example, the opening of a large industrial plant) poses excessive risks to human health and, if so, what protective measures are necessary to mitigate these risks, Also based on a balancing judgement between the needs of economic development and the fundamental protection of health.

One aspect to be taken into account, when adhering to the *One Health* logic, is that of the relationships between the different territorial levels of government in the Italian constitutional system. In fact, considering the problems related to human health, animals and the environment means implementing policies that involve, across sectors attributed both to legislative and regulatory competence of the State and, in part, of the Regions. As a result, coordination mechanisms at state and regional level are needed to ensure a satisfactory level of cooperation between heterogeneous actors.

As mentioned above,[73] the subjects involved in the *One Health* paradigm follow a different allocation of competences between State and Regions. In fact, according to art. 117 The protection of health is assigned to the concurrent legislative competence of the State and the Regions,[74] while the protection of the environment and animals belongs exclusively to the State.[75]

Therefore, if the highest possible levels of public health protection are to be achieved in the future, *One Health*, State and Regions must find tools and forms of collaboration that go beyond their different formal competences.[76] Otherwise, it would mean that the different allocation of competence for closely related matters, such as health and environment, represents an obstacle to increasing the level of overall protection offered by the Italian law.

Similar doubts arise from the attribution to the autonomous regions and provinces, by decree of the Ministry of Health of 9 June 2022, the task of establishing a regional system for prevention of environmental and climate risks, which contributes to the achievement of primary prevention objectives of National Health Prevention System. On the contrary, the regional replication of the system introduced by the Decree Law No. 36/2022 at the state level shows how the ambition of One Health to treat in an integrated manner sectors attributed to different territorial authorities compels to design a suitable system coordination and cooperation.

6 Conclusions

From the reflections carried out, emerge two critical aspects: the first concerns the level of international and European law, while the other strictly concerns the Italian legal system.

Regarding the supranational dimension, although the regulatory acts adopted by the Tripartite+ organizations have had the important merit of defining the *One Health* paradigm, requiring States to implement it in their respective systems, the set of rules introduced belong to the category of soft law.

---

[73] Cf. Sect. 4.3.

[74] In any case, Article 117 Cost. states that it is the State's responsibility to determine the essential levels of assistance.

[75] Cf. Delsignore et al. (2022), p. 20.

[76] Cf. Fracchia (2022), p. 22.

From this, two observations follow. First, national states are not obliged to implement the international prevention strategy inspired by the *One Health* model. Secondly, States retain full discretion also about the methods of implementation of this approach and the potential areas of its application. All this, however, seems to represent an obstacle to the realization of the essential objective of the *One Health* approach: the protection of health at global level.

The problems just highlighted could disappear if there were a high rate of spontaneous adherence to the regulatory acts of international organizations; in this case, the One Health principles could in fact be included among the sources of customary international law.[77]

Furthermore, European public policies on *One Health* are contained in acts that are not binding on the Member States. An exception appears to derive from EU Regulation 522/2021, which, however, limits itself to providing for an allocation of financial resources, together with the related spending and reporting methods, for the implementation of the European Health Program 2021–2027.

As regards the Italian case, the main problem concerns the forms of coordination at a horizontal and vertical level.

As we have seen, the state legislator, despite involving various institutional subjects within the National Health Prevention System (Ministry of Health, Regions, Autonomous Provinces and health authorities), has not defined relevant cooperation tools.

In addition, this criticality also arises with regard to the coordination between the National System and the Regional Health Prevention Systems, which are assigned the task of contributing to primary prevention at regional level.

In the end, the Italian legislator has followed up on the guidelines adopted at a supranational and European level, although difficulties remain regarding the concrete creation of a healthcare system fully corresponding to the holistic vision of health, as required by the *One Health* approach.

References

Amirante D (2022) Costituzionalismo ambientale. Atlante giuridico per l'Antropocene. Mulino, Bologna

Bifulco R (2022) Primissime riflessioni intorno alla l. cost. 1/2022. Federalismi.it 11:1–8

Black PF, Butler CD (2014) One Health in a world with climate change. Rev Sci Technol 33(2):465–473. https://doi.org/10.20506/rst.33.2.2293

Boissons De Chazournes L, Maljean Du Bois S (2016) Principes de droit international de l'environnement. Jurisclasseur environnement et development durable 1:1–25

Caravita B, Morrone A (2016) Ambiente e Costituzione. In: Caravita B, Cassetti L, Morrone A (eds) Diritto dell'ambiente. Mulino, Bologna, pp 17–38

Connolly J (2017) Governing towards 'One Health': establishing knowledge integration in global health security governance. Glob Policy 4:483–494. https://doi.org/10.1111/1758-5899.12505

Costato L, Manservisi S (2012) Profili di diritto ambientale nell'Unione europea. Cedam, Padova

---

[77] See para 2.

D'Atena A (1974) L'autonomia legislativa delle Regioni. Bulzoni, Roma
De Grove-Valdeyron N (2018) Droit européen de la santé. Lextenso éditions, Paris
Delsignore M, Marra A, Ramjoli M (2022) La riforma costituzionale e il nuovo volto del legislatore nella tutela dell'ambiente. Rivista giuridica dell'ambiente 1:1–38
Di Federico G, Negri S (2020) Unione europea e salute. Principi, azioni, diritti e sicurezza. Cedam, Padova
Di Salvatore E (2015) Le politiche europee nella spirale della crisi economica e l'impatto sull'organizzazione e sul funzionamento dello Stato membro. In: Di Salvatore E (ed) L'Europea e noi. Galaad Edizioni, Giulianova
Di Salvatore E (2022) Brevi osservazioni sulla revisione degli articoli 9 e 41 della Costituzione. Costituzionalismo.it 1:1–22
Di Salvatore E (2023) Interpretazione e nozioni della Costituzione. Costituzionalismo.it 1:1–26
Di Salvatore E (2024) L'evoluzione dei fini e dei compiti dello Stato e il dovere di protezione dei diritti fondamentali. Rivista "Gruppo di Pisa" 1:93–107
Flaminia AB (2022) Una lezione conclusiva che vuole essere un punto di partenza. In: Flaminia AB (ed) One Health: la tutela della salute. Editoriale Scientifica, Napoli, pp 9–21
Fracchia F (2022) L'ambiente nell'art. 9 della Costituzione: un approccio in "negativo". Il diritto dell'economia 1:15–30
Galicz KZ (2022) One Health, one society: il diritto alla salute quale presupposto per il diritto di asilo. Corti supreme e salute 3:753–777
Garcia SN, Osburn BI, Jay-Russell MT (2020) One Health for food safety, food security, and sustainable food production. Front Sustain Food Syst 4:1. https://doi.org/10.3389/fsufs.2020.00001
Gibbs P (2014) The evolution of One Health: a decade of progress and challenges for the future. Vet Rec 4:85–91. https://doi.org/10.1136/vr.g143
Iacovino A, Randazzo L (2023) Climate change and food security in the One Health perspective. BioLaw J - Rivista di BioDiritto 2:293–315
Karesh W, Cook RA (2005) The human-animal link, one world - One Health. Foreign Aff 84:38–50
Kramer L (2016) EC environmental law. Bloomsbury Publishing, Londra
Latino A (2022) Il paradigma One Health nell'ordinamento internazionale: un'analisi critica di origini, protagonisti, strumenti normativi. Corti Supreme e Salute 3:779–808
Lerner H, Berg C (2017) A comparison of three holistic approaches to health: One Health, EcoHealth, and Planetary Health. Front Vet Sci 4:163. https://doi.org/10.3389/fvets.2017.00163
Luciani M (1991) Diritto alla salute - Dir. cost. Enciclopedia giuridica XXXII:1–14
Mangiameli S (1986) La proprietà privata nella Costituzione italiana. Profili generali. Giuffrè, Milano
Morana D (2002) La salute nella Costituzione italiana. Profili sistematici. Giuffrè, Milano
Pace A (2003) Problematica delle libertà costituzionali. Cedam, Padova
Pajno S, Pucci V (2020) Il diritto "fondamentale" alla salute nei procedimenti di valutazione ambientale. In: Di Folco M, Mengozzi M (eds) La salute nelle valutazioni di impatto ambientale. Edizioni Scientifiche Italiane, Napoli, pp 2–62
Pezzini B (1983) Il diritto alla salute: profili costituzionali. Diritto e società 1:21–91
Ragone G (2022) *One Health* e Costituzione italiana, tra spinte eco-centriche e nuove prospettive di tutela della salute umana, ambientale e animale. Corti supreme e salute 3:809–826
Schwabe CW (1984) Veterinary medicine and human health. Williams & Wilkins, Baltimore
Vipiana P (2022) La protezione degli animali nel nuovo art. 9 Cost. DPCE 2:1111–1121
Zinsstag J, Schelling E, Waltner-Toews D, Tanner M (2011) From "one medicine" to "One Health" and systemic approaches to health and well-being. Prev Vet Med 101(3-4):148–156. https://doi.org/10.1016/j.prevetmed.2010.07.003

The Sustainable Use of Healthy Soil as a Cornerstone of the One Health Approach: An Analysis of European Public Strategies and Market-Based Environmental Protection Tools

Alessia Depietri

Abstract Soil, being a non-renewable resource, faces increasing pressure due to the burgeoning global population and its concomitant exploitation. This scenario presents a foreseeable global challenge characterized by resource scarcity. In response, the European Commission has introduced new strategic guidelines, underscoring the critical role of sustainable soil management as a prerequisite for achieving the climate and biodiversity targets set forth in the European Green Deal and the United Nations Agenda 2030. In this framework, the One Health approach is particularly pertinent. This paper aims to highlight that protecting soil health—an integral element of a comprehensive health paradigm—requires the implementation of public policies that safeguard biodiversity, ensure nutritional security, and uphold both human and animal health. These policies must also address additional goals such as enhancing employment and market competitiveness. Particular emphasis should be placed on the enforcement of the Common Agricultural Policy (CAP) for 2023–2027, the commencement of CAP strategic plans across the 27 EU member states, and the pivotal role of both European and national bioeconomy strategies. Moreover, it is essential to examine how legal instruments—whether hard law or soft law—can function as enablers rather than impediments in this endeavor. When effectively structured, market-based environmental protection instruments can play a significant role in promoting sustainable agricultural practices among producers and consumers alike.

A. Depietri (✉)
University of Parma, Parma, Italy
e-mail: alessia.depietri@unipr.it

G. Antonelli et al. (eds.), *Planetary Health - Laws, Policies and Science on the 'One Health' Approach*, https://doi.org/10.1007/978-3-031-90621-3_2

1 Introduction

If one word were chosen to describe the current global scenario, *polycrisis* would be the most fitting. In social sciences,[1] *polycrisis* refers to a complex situation where interconnected problems, antagonisms, and uncontrolled processes converge. The World Economic Forum's "*Global Risks Report 2023*"[2] describes it as a phenomenon where multiple present and future risks interact, compounding their effects, and leading to an overall impact greater than the sum of its parts.

From this definition, we can identify key concepts that serve as a foundation for this essay: risks, interaction, effects, and impact.

Risk refers to the potential for harm, and when risks of different natures interact, their effects are amplified, intensifying the severity of the resulting impacts.

This concise formula applies to our reality, where the intersection of current events and the looming risks of health, food, and ecological catastrophes underpin the global challenge we face, fueled by the hope of avoiding a point of no return.

It's as if the world were a fragile puzzle, where the loss of each piece gradually erases the original design, disrupting its balance.

This is the challenge for policymakers: to piece together a crumbling puzzle so that the human-nature balance can be restored.

International, European and national institutions have taken on this challenge, anchoring their efforts in the vital role of public policy as a catalyst for change in both the present and future. In these contexts, the principle of sustainable development[3] has become central to aligning the political agendas of various member states

[1] The concept of *polycrisis* originates from the studies of philosophers of complexity, particularly those of Edgar Morin and Anne Brigitte Kern. They explore the idea of multiple, interconnected crises affecting the world simultaneously. The term *polycrisis* reflects the complex and interdependent nature of these crises, where each issue is intertwined with others, creating a cumulative and often unpredictable impact on global systems. For more on these aspects, one can refer to Morin and Kern (1999).

[2] The World Economic Forum's "*Global Risks Report 2023*" outlines the major global risks expected to impact the world over the next decade. Key themes of the report include the Cost of Living Crisis, Geopolitical Tensions, Climate Change, Cybersecurity and Digital Risks, Societal Challenges. The report aims to help policymakers and leaders understand these risks and develop strategies to build global resilience and prepare for future challenges. The "Global Risks Report 2023" from the World Economic Forum can be downloaded for free from the official World Economic Forum website. On the site, users can access the full report, which includes detailed analyses of global risks and recommendations for leaders and policymakers.

[3] On the topic of sustainable development, the definition provided in the Brundtland Report of 1987 is well-known. It defines sustainable development as "development that meets the needs of the present generation without compromising the ability of future generations to meet their own needs" (World Commission on Environment and Development, *Our Common Future*, Oxford Paperbacks, 1990).

In terms of scholarly literature, key works include Rajamani and Peel (2021); Roeben (2020); Sands et al. (2018); Sachs (2014); Fracchia (2010); Bugge and Voigt (2008); Schrijver and Weiss (2004).

for the well-being of current and future generations and the protection of the ecosystem.

For instance, the framework of the *European Green Deal*[4] and the *United Nations' 2030 Agenda*[5] places the responsibility on current generations to "think about the future", emphasizing that sustainability and resilience are the guiding principles for international, European and national legislative actions.

A key example is the goal of climate neutrality, a cornerstone of the *European Green Deal*, implemented through the *European Climate Law*[6] and the *Fit for 55 package*.[7]

Furthermore, *The Next Generation EU*[8] initiative (designed to support European states in their post-pandemic recovery) also prioritizes financing and incentivizing a green and digital transition, with a focus on biodiversity protection, research promotion, and alternative production models.

Early examples of recognizing new sustainability needs can also be found in various European constitutions.[9]

---

[4] The European Green Deal was adopted through a Communication from the European Commission to the European Parliament, the Council, the European Economic and Social Committee, and the Committee of the Regions on December 11, 2019 (COM) 2019 640 final. This document outlines the initiatives that the Commission will present as part of its Work Programme to implement a fair and inclusive transition towards a sustainable economy in the European Union. The Communication emphasizes the need for a new pact that brings together citizens, national, regional, and local authorities, civil society, and industry, working closely with EU institutions and advisory bodies. For further reading on the Green Deal, consider the following literature: Bevilacqua and Chiti (2024); Chomsky and Pollin (2020); Pettifor (2019).

[5] *Agenda 2030*, adopted by the UN in 2015, sets out 17 *Sustainable Development Goals (SDGs)* to tackle global challenges like poverty, inequality, and climate change by 2030. It promotes balanced progress across economic growth, social inclusion, and environmental protection, calling for global cooperation and local action to ensure a sustainable and equitable future for all. For this topic, please refer to Beisheim et al. (2022), pp. 22 ff; Novitz and Pieraccini (2020), pp. 39 ff.

[6] Regulation (EU) 2021/1119: European Climate Law. It sets the goal for Europe to become climate-neutral by 2050 and targets a 55% reduction in greenhouse gas emissions by 2030 compared to 1990 levels. To achieve climate neutrality, the EU aims for net zero emissions through emission cuts, green technology investments, and environmental protection. The law ensures all EU policies and sectors contribute to these objectives

[7] For further details, see Communication COM (2021) 550 final, dated July 14, 2021, titled *Fit for 55: Achieving the EU's 2030 Climate Target on the Path to Climate Neutrality*. This package outlines new targets for renewable energy and transportation emissions and introduces a social climate fund financed by ecological taxes.

[8] *NextGenEU* is the EU's €800 billion recovery plan to address COVID-19's impact, focusing on green and digital transitions, economic recovery, and resilience. It provides funding to member states for sustainable growth and job creation. In particular, Regulation (EU) 2021/241 (establishing the Recovery and Resilience Facility) stipulates those only measures complying with the *Do No Significant Harm* (DNSH) principle can be funded within the national plans.

[9] The preamble of the French Constitution, for instance, references the duties outlined in the Environmental Charter, which integrates principles guaranteeing the protection of the ecosystem, human health, and future generations, highlighting their close interrelation. The Finnish Constitution, in Article 20, emphasizes that nature, biodiversity, the environment, and national

There is no doubt that policymakers grasp the seriousness of the current situation. The emphasis now should not be on identifying critical issues and areas for intervention, but rather on interconnecting the various sustainability needs through an interdisciplinary strategy, one that is humanistic but not anthropocentric and rooted in ethical scientific principles.[10] In other words, tackling the *polycrisis* of the Anthropocene requires public policy to embrace a global approach that redefines humans as part of an ecosystem, rather than its central protagonists.

The approach being discussed is the One Health approach,[11] which addresses the critical need to integrate human, animal, and ecosystem health[12] into a cohesive and interconnected system.[13]

In other words, it aims to improve overall health outcomes by integrating these areas and fostering collaboration across disciplines. This approach acknowledges that changes or issues in one domain can affect the others, and it seeks to address health challenges more effectively through a comprehensive, cross-sectoral strategy.

---

heritage are the responsibility of everyone. Public authorities are committed to ensuring the right to a healthy environment and the opportunity for public participation in decisions affecting it.

The 2022 Italian Constitutional reform, by amending Articles 9 and 41, introduced the principles of environmental protection, biodiversity, and ecosystems into the constitutional framework, emphasizing the interests of future generations. This reform defines the environment, health, and ecological purposes as fundamental goals of economic activity, challenging the notion of incompatibility between environmental protection and the market, and underscoring the need for sustainable and ecological economic progress in line with the redefined European development model.

[10] These aspects inevitably tie back to the broader catalyst of bioethics, as described in an almost prophetic manner in Van Potter (1971).

[11] To clarify from the outset, the One Health approach is not the sole holistic framework in use. The literature indicates that several integrated normative approaches are recognized in environmental matters, with broad support from both the scientific community and international institutions. Besides One Health, these include the Ecosystemic approach, the Ecological approach, and the Human Rights-Based (HRB) approach. For a general overview, see Pitto (2023), pp. 315 ff. Upon brief reflection, it is important to highlight that the One Health approach and agroecology—especially its emphasis on collaborating with natural systems to achieve sustainable food production—can be seen as complementary in the quest for a world where all individuals have access to safe and nutritious food while also protecting the environment. The FAO is actively working to promote both agroecology and One Health through its various programs and initiatives and assists countries in adopting these approaches. For a comprehensive and updated review of agroecology from a legal standpoint, please consult Monteduro (2015), pp. 57 ff.

[12] About these connections, see Rabinowitz and Conti (2013), pp. 189 ff.

[13] About the importance of this holistic approach, see Mackenzie and Jeggo (2019), pp. 88 ff.; Sweeney et al. (2018), pp. 34 ff.; Kazwala et al. (2017), pp. 1725 ff.; Cleaveland et al. (2018), pp. 193 ss; Atlas (2013), pp. 1 ff.; Daszak et al. (2001), pp. 103 ff.

2 The One Health Approach in Summary: Milestones and Significance

The evolution[14] of the One Health approach has closely followed major global pandemics.

In September 2004, amid the SARS outbreak, a group of strategic thinkers and health experts gathered for the symposium "*Building Interdisciplinary Bridges to Health in a Globalized World*", organized by the Wildlife Conservation Society. This event focused on the connections between human, domestic animal, and wildlife diseases. The symposium led to the creation of the "*Manhattan Principles on One World, One Health*",[15] which outlined twelve recommendations aimed at preventing epidemics and epizootics and preserving ecosystem integrity for the benefit of humans, animals, and biodiversity. The core idea was that understanding health and disease requires an integrated approach that spans different disciplines and sectors, addressing prevention, surveillance, monitoring, and environmental conservation. The rise of infectious diseases not only threatens human health and economies—particularly food systems[16]—but also impacts flora and fauna, essential for maintaining global biodiversity.

The outcomes of this conference laid the groundwork for the 2008 strategic framework adopted by FAO, OIE, and WHO, in collaboration with UNICEF, the United Nations System Influenza Coordination (UNSIC), and the International Bank for Reconstruction and Development (IBRD). This framework aimed to coordinate global responses to emerging and re-emerging diseases. The One Health approach then transitioned from a recommendation by experts to a key focus for various international health organizations.

---

[14] For a comprehensive overview of the topic, covering its main stages of development, please refer to Scotti (2023), pp. 45 ff.

[15] Here is the preamble to the 12 principles, "Recent outbreaks of West Nile Virus, Ebola Hemorrhagic Fever, SARS, Monkeypox, Mad Cow Disease and Avian Influenza remind us that human and animal health are intimately connected. A broader understanding of health and disease demands a unity of approach achievable only through a consilience of human, domestic animal and wildlife health—One Health. Phenomena such as species loss, habitat degradation, pollution, invasive alien species, and global climate change are fundamentally altering life on our planet from terrestrial wilderness and ocean depths to the most densely populated cities. The rise of emerging and resurging infectious diseases threatens not only humans (and their food supplies and economies), but also the fauna and flora comprising the critically needed biodiversity that supports the living infrastructure of our world. The earnestness and effectiveness of humankind's environmental stewardship and our future health have never been more clearly linked. To win the disease battles of the 21st Century while ensuring the biological integrity of the Earth for future generations requires interdisciplinary and cross-sectoral approaches to disease prevention, surveillance, monitoring, control and mitigation as well as to environmental conservation more broadly". Cook et al. (2004), p. 29.

[16] Food sustainability is undoubtedly one of the most pressing issues in the context of effectively implementing the One Health approach. For relevant contributions in the literature, see, for example Ray et al. (2019); Berry et al. (2015), pp. 2293 ff.; Tilman and Clark (2014), pp. 518 ff.

The 2010 Hanoi Declaration further advocated for its widespread application. The formal recognition of the One Health model is largely due to the collaborative efforts of FAO, OIE, and WHO, working together under the Tripartite Alliance. This partnership has moved the One Health concept from theory to practice, implementing concrete actions.

In 2017, the third session of the United Nations Environment Assembly emphasized the need to address the links between environment and health together as part of the 2030 Agenda for Sustainable Development.

In 2021, following the pandemic's onset, FAO, OIE, WHO, and the United Nations Environment Programme (UNEP) established a high-level One Health expert group. This group aims to gather and share scientific data on the interactions between environment, human health, and animal health to better handle future health crises.

Overall, COVID-19 has highlighted that no single government or institution can tackle the threat of future pandemics alone.[17] As infectious diseases increase and cross borders, it is crucial to be better prepared and more resilient. Therefore, the 194 WHO member countries are collaborating on a new global instrument designed to protect people, communities, and nations from future pandemics, with the One Health strategy being a key component of this effort.

The deeper significance of the One Health approach involves addressing multiple dimensions of health and balancing the values and interests associated with the human, animal, and environmental pillars.

In the legal context, this represents a shift in environmental policies and the tools used to implement them. However, as some scholars have observed,[18] the One Health approach has not always managed to move beyond an anthropocentric perspective. In practice, the animal and environmental sectors are often still viewed primarily through their impact on human well-being, whether as risks or benefits, which can undermine the effectiveness of the strategy.

In this context, a particularly important area of investigation is the environmental policies related to soil protection, especially within the European Union. The aim is to determine whether these policies are supported by effective tools and whether they achieve significant results in advancing the One Health approach as part of a sustainable economy.

Indeed, soil plays a crucial role in connecting the interests of humans, animals, and the environment. Its health is fundamental to the One Health strategy: it affects human health, nutrition, and economic well-being, supports wildlife, and maintains the delicate balance of biodiversity. Thus, soil health can be seen as a cornerstone of this global strategy.[19]

---

[17] On this point, see Musa et al. (2020), pp. 244 ff.

[18] The reference is to Keune et al. (2022), pp. 93 ff; Kamenshchikova et al. (2019), pp. 306 ff; Craddock and Hinchliffe (2015), pp. 1 ff.; Wallace et al. (2015), pp. 68 ff.

[19] On this point, see Lal et al. (2021), e00398.

3 The Importance of Soil Sustainability: Data, Scientific Evidence, and Policy Insights

As science shows,[20] soil is a dynamic and variable natural resource made up of minerals, water, air, and organic matter. Its unique properties, resulting from rock weathering, organic decomposition, and biological and chemical processes, allow it to support plant growth, regulate water flow, and host a variety of organisms. Healthy soils are the planet's largest carbon sink and also have the ability to absorb water, which helps to reduce the risk of floods and droughts. To emphasize the importance of soil for the entire ecosystem, it's worth noting that it contains over 25% of the planet's total biodiversity[21] and it is essential for the entire food chain.[22]

However, the data shows a worrying trend: it is estimated that 60% to 70% of soils in the EU are degraded. Scientists describe this degradation as either the complete loss of soil or the loss of its ability to support the biological communities that typically live there.[23]

It's clear that, in most cases, soil degradation is due to improper human activities (think of human-caused issues like excessive erosion, salinization, urbanization, pollution, desertification, and soil sealing).[24]

These percentages are even more concerning when viewed alongside the steady growth of the global population. In fact, the data indicates a situation of overpopulation that could reach nearly 10 billion people by 2050.[25] Along with the environmental damage, there is also an economic cost, estimated to exceed 50 billion euros annually.[26]

In short, soil and its resources are non-renewable. As the global population grows, the excessive and improper use of soil is likely to lead to a future where essential resources become increasingly scarce.

For these reasons, the public policy sector can play a key role in reversing this trend.[27]

---

[20] For a comprehensive overview of the significance of soil from a scientific standpoint, refer to by Brady and Weil (2007); Montgomery (2007), pp. 13268 ff.; Brady, N.C. and Weil, R.R. (2008) *The nature and properties of soils* 14 ed. Prentice-Hall, Upper Saddle River, New Jersey.. Prentice and Ohlson (2014).

[21] About this aspect refer to FAO (2020).

[22] See European Commission, Directorate-General for Research and Innovation, Veerman et al. (2020).

[23] *Ibidem.*

[24] On the failures of human soil management, see Blaikie and Brookfield (2015).

[25] On this point see World Resources Institute (2019) *Creating a sustainable food future.*

[26] European Commission, Directorate-General for Research and Innovation, Veerman et al. (2020). With reference to the costs of land degradation see Nkonya et al. (2016).

[27] Blaikie (2016); Glæsner et al. (2014), pp. 9538 ff.

As European institutions have pointed out, maintaining healthy soils is vital for achieving the goals of the *Green Deal* related to climate, biodiversity, and sustainable economic development.[28]

The EU Soil Strategy for 2030[29] fits within this context. The subtitle, "*Reaping the Benefits of Healthy Soils for People, Food, Nature, and Climate*", suggests a holistic approach as its foundation.

The strategy's purpose is immediately clear: to uphold the core principles of the One Health approach, which it explicitly mentions.

The strategy is notable for its strong emphasis on both interdisciplinarity and interconnectedness. It draws on extensive references and concrete data from natural science experts, highlighting its interdisciplinary nature. As for interconnectedness, the strategy is closely aligned with other EU initiatives,[30] including those on the *Bioeconomy* and the *Farm to Fork strategy.*[31]

Furthermore, to support the transition towards healthy soils by 2030, the Mission *A Soil Deal for Europe* will allocate funding for a range of research and intervention initiatives.[32]

To tackle the cross-border effects of soil degradation, ensure fair market conditions, and improve alignment between EU and national policies on climate change, biodiversity, food security, and water protection, the strategy emphasizes the need for new, unified legal instruments among member states.

---

[28] See De Leonardis (2022), pp. 400 ff; Panagos et al. (2018), pp. 383 ff.; Mcbratney et al. (2013), pp. 203 ff.

[29] See the Communication from the Commission to the European Parliament, the Council, the European Economic and Social committee and the Committee of the Regions EU, *Soil Strategy for 2030 Reaping the benefits of healthy soils for people, food, nature and climate*, COM (2021) 699 final. On page 4, the Communication defines its core focus «To address transboundary impacts of soil degradation, secure equal market conditions, promote policy coherence at EU and national level and thus be able to achieve our goals on climate change, biodiversity, food security and water protection, the Commission will table a dedicated legislative proposal on soil health by 2023 which will enable the objectives of this strategy to be met and good soil health to be achieved across the EU by 2050. Such a legislative initiative will fulfil better regulation requirements, be based on a thorough impact assessment, including a subsidiarity check, and fully respect the competences of Member States in this matter. To determine the scope and content of this proportionate and risk-based framework, the Commission will engage in a broad and inclusive consultation with Member States, the European Parliament and all relevant stakeholders».

[30] In particular, see the *EU Biodiversity Strategy for 2030*, COM (2020)380 and the *EU Climate Adaptation Strategy*, COM (2021) 82.

[31] The decision to explicitly reference the *European Bioeconomy Strategy* and the *Farm to Fork Strategy* is intended to examine the extent to which the One Health approach is incorporated within the agricultural sector, which, by its very nature, is most intimately connected with the soil.

[32] Its main goal of the Mission is to establish 100 living labs and lighthouses to lead the transition towards healthy soils by 2030, leading the transition towards healthy soils by funding an ambitious research and innovation programme, developing a harmonised framework for soil monitoring in Europe and raising people's awareness on the vital importance of soils.

These should replace or enhance the existing EU soil protection laws,[33] and the various measures adopted under the 2006 *Thematic Strategy for Soil Protection,*[34] which are seen as insufficient.

Closely linked to the *Soil Strategy* is the European Parliament and Council's proposed directive on *Soil Monitoring and Resilience,*[35] which aims to achieve healthy soils across the EU by 2050. This directive focuses on measures to monitor and assess soil health, manage soil sustainably, and address contaminated sites by defining, identifying, and evaluating their risks.

The monitoring will involve assessing a set of indicators at least every 5 years to track soil degradation. These indicators will be connected to specific evaluation criteria, with some set at the European level and others determined by individual member states.

Member states will need to report on soil health monitoring and assessment every 5 years to the European Commission and the European Environment Agency (EEA), which will make this data available through a European portal.

However, the proposed directive, now scaled back to a *Soil Monitoring Law*, appears significantly less ambitious than the original *Soil Strategy for 2030* which aimed to establish a comprehensive *Soil Health Law*. This reduction underscores the need for European and national legal frameworks to develop more robust tools where the One Health approach is fully integrated, rather than just a theoretical concept.

As mentioned above, among the various strategies developed by the European Union, the importance of soil health is addressed across several areas, including the *Farm to Fork strategy* and the strategies on the *Bioeconomy*.

The *Farm to Fork strategy*[36] is focused on creating sustainable food systems by considering every stage of the food chain. It aims to establish an integrated agri-food policy, rather than one divided among traditional sectors like agriculture, environment, health, trade, and energy, to prevent uncoordinated rules and actions.

Essentially, the strategy seeks to transform the role of food systems from being major contributors to climate and environmental issues to serving as a potential key solution for achieving climate neutrality and significantly boosting the economic competitiveness of EU countries in this sector.[37] This strategy is undoubtedly

---

[33] The reference pertains to certain directives that address specific soil characteristics, albeit in a somewhat fragmented manner. See, for example, the Industrial Emissions Directive 2010/75/EU, the Environmental Liability Directive 2004/35/CE, the LULUCF Regulation 2018/841/EU and the Sewage Sludge Directive 86/278/CEE.

[34] Commission Communication of 22 September 2006 entitled "*Thematic strategy for soil protection*" COM (2006) 231 final.

[35] See the Proposal for a Directive of the European Parliament and of the council on Soil Monitoring and Resilience (Soil Monitoring Law) COM (2023) 416 final. Generally, for insights on the importance of soil management, see Helming et al. (2018), pp. 4432 ff.

[36] See the Communication from the Commission to the European Parliament, the Council, the European economic and Social Committee and the Committee of the regions, *A Farm to Fork Strategy for a fair, healthy and environmentally-friendly food system,* COM (2020) 381 final.

[37] Schebesta and Chandel (2020), pp. 586 ff.; Venturi (2021), pp. 70 ff.

inspired by the One Health approach. It is a key component of the *Biodiversity Strategy* and aims to develop resilient and safe agri-food systems for humans while preserving animal welfare and ecosystem integrity. The success of this strategy depends on the sustainable use of soil, particularly in agricultural practices.

As a result, this strategy sets the stage for advancing organic farming, as detailed in the related action plan.

Additionally, consider the conclusions adopted by the European Council on carbon sequestration in agricultural soils. These conclusions align perfectly with the European Commission's December 2021 communication on sustainable carbon cycles.[38] They aim to promote agricultural practices that capture carbon from the atmosphere and store it sustainably in soils or biomass.[39] Examples include planting hedges or trees, cultivating legumes, and using cover crops and intercropping.

All these aspects must be viewed also in the context of the European *Bioeconomy strategy,*[40] which was first established in 2012 and significantly updated in 2018.

The strategy focuses on five main goals: ensuring food and nutritional security, managing resources sustainably, reducing reliance on non-renewable and unsustainable resources, mitigating and adapting to climate change, and boosting European competitiveness and employment.

To promote soil and biodiversity regeneration, the bioeconomy harnesses the potential of biological resources in the agro-industrial sectors, aiming for sustainable and circular[41] economic development.[42]

To achieve these goals, three key action areas are crucial. First, the biological sectors need to be strengthened and expanded, including unlocking investments and market opportunities. Additionally, bioeconomic strategies should be promoted at

---

[38] On this point see the Communication from the Commission to the European Parliament and the Council, *Sustainable Carbon Cycles,* COM (2021) 800 final.

[39] This aspect is undoubtedly the outcome of extensive scientific research into the potential of soils as carbon sinks. For a comprehensive overview, please refer to: Lal (2004), pp. 1 ff. Olofsson and Börjesson (2018), pp. 997 ff.; Forfora et al. (2024), 140677.

[40] On this point see European Commission, Directorate-General for Research and Innovation 2018, *A Sustainable Bioeconomy for Europe: Strengthening the Connection Between Economy, Society, and the Environment – Updated Bioeconomy Strategy,* Publications Office, ID. (2018), *Review of the 2012 European Bioeconomy Strategy,* Publications Office.

[41] With particular reference to the interconnection between the circular economy and the bioeconomy, see De Leonardis (2022), pp. 110 ff. Among the various initiatives in this domain, it is noteworthy that the Circular Bioeconomy Partnership (CBE) for the period 2021–2031, established under Regulation (EU) 2021/2085, has been allocated an EU contribution of €1 billion. This funding is intended to progressively strengthen and expand the EU's bioeconomy sectors throughout all stages of the innovation cycle. Additionally, the recent establishment of the European Fund for Circular Bioeconomy is significant. This fund represents the first dedicated risk fund in Europe specifically focused on the bioeconomy and circular bioeconomy, providing financing through Horizon 2020 and the European Investment Bank.

[42] In the academic literature, contributions include Bell et al. (2018), pp. 25 ff.; Philp (2018), pp. 11 ff.; Sanz-Hernández et al. (2019), pp. 107 ff. Vivien et al. (2019), pp. 189 ff.

the local level within member states, and the ecological limits of the bioeconomy must be clearly understood.[43]

Recently, the European Commission released a report on the progress of this strategy, showing overall positive developments. However, it points out the need for better management of land and biomass demands and for enhancing sustainable consumption models.

The report reiterates that effective soil management is essential for meeting the EU's environmental goals.[44]

Unlike the clearly holistic approach of the *Farm to Fork strategy*, the bioeconomy strategy still displays a lot of anthropocentric elements. The strategy focuses on the sustainable use of nature primarily in terms of human well-being, especially from an economic perspective. This does not mean that the bioeconomy strategy is environmentally negative. However, it highlights the need for a more cohesive approach by European institutions in addressing the different aspects of public policy.

In summary, the array of interconnected European strategies for ecological transition fundamentally relies on preserving soil health, integrating scientific evidence about its current condition and the improvements needed in its management.

Regarding the One Health approach, two key points emerge: first, the holistic approach is applied in some strategies but not others, which still maintain an anthropocentric perspective; second, there is a pressing need to translate the principles of this model from strategic planning into practical, actionable measures.

4 The Common Agricultural Policy (CAP): Connections and Disconnections Through the Lens of the One Health Approach

Focusing on the application of the One Health framework to soil sustainability, it is crucial to emphasize the environmental policies tools strictly related to agriculture, which is the human activity that most significantly impacts soil health.[45]

The focus must be on the Common Agricultural Policy (CAP), which was initially established in 1962 and has been continuously updated since.[46]

---

[43] To this end, the European Commission has established several working groups with Member States to facilitate the swift development of local bioeconomies across Europe. The European Bioeconomy Policy Forum supports the exchange of knowledge and best practices among Member States, aiding in the development and implementation of national and regional bioeconomy policies.

[44] Report COM (2022) 283, *EU Bioeconomy Strategy Progress Report - European Bioeconomy policy: stocktaking and future developments.*

[45] For a global perspective on this issue, see Faye and Braun (2022), 102799.

[46] When the common market was established by the Treaty of Rome in 1958, the agricultural sectors of the six founding Member States were heavily influenced by state intervention. To integrate

The CAP represents a close partnership between agriculture and society, linking Europe with its farmers; it is managed and funded at the European level through the EU budget.[47]

The CAP for 2023–2027,[48] which came into effect on January 1, 2023, includes various strategic plans aimed at supporting the European Green Deal's goals and its measures for biodiversity[49] and sustainable agri-food production and consumption.

From an operational standpoint, the main role of the CAP is to support agriculture for several reasons.[50] Agriculture is crucial for human sustenance but is also highly unpredictable due to its reliance on weather conditions. On average across Europe, agriculture is less profitable, with agricultural incomes being about 40% lower than non-agricultural incomes.

To address these issues, the CAP employs various tools, including income support funds, market interventions, and rural development measures.

From a strictly legal standpoint, the CAP is a crucial area of EU regulation. Its primary legal basis is found in Articles 38–44 of the Treaty on the Functioning of the European Union (TFEU).

For the purposes of this discussion, it is enough to say that Article 38(1) states that "*The Union shall define and implement a common agricultural and fisheries policy*," while Article 39 outlines its specific objectives.[51]

---

agricultural products into the free movement of goods while maintaining public intervention in agriculture, it was necessary to eliminate national intervention mechanisms that were incompatible with the common market. These mechanisms were then adapted to the community level, which ultimately led to the creation of the Common Agricultural Policy (CAP).

[47] The Multiannual Financial Framework (MFF) of the European Union for the 2021–2027 period, which was approved on December 17, 2020, allocates €1.21 trillion (in current prices). In addition, the NextGenerationEU recovery plan contributes a further €808 billion. The total funding dedicated to the Common Agricultural Policy (CAP) is €386.6 billion, which is split between two funds, commonly known as the "two pillars" of the CAP. The European Agricultural Guarantee Fund (EAGF), known as the "first pillar" of the CAP, is allocated a total of €291.1 billion. Of this sum, up to €270 billion will be used for income support schemes, while the remainder will be allocated to agricultural market support initiatives. The European Agricultural Fund for Rural Development (EAFRD), known as the "second pillar" of the CAP, has a total allocation of €95.5 billion, including €8.1 billion from NextGenerationEU. This recovery fund, designed to address the COVID-19 pandemic's challenges, allocated 30% of its resources in 2021 and released the remaining 70% in 2022.

[48] Röder et al. (2024), 107219. The article examines the final versions of the legal texts for both the current and previous CAP to evaluate whether there is potential for enhanced environmental ambition.

[49] Pe'Er et al. (2022), e12901.

[50] Pe'Er et al. (2019), pp. 449 ff.

[51] Article 39 of the Treaty on the Functioning of the European Union (TFEU) sets out the objectives of the Common Agricultural Policy (CAP). The text reads as follows: 1. The objectives of the common agricultural policy shall be: '(a) to increase agricultural productivity by promoting technical progress and by ensuring the rational development of agricultural production and the optimum utilization of the factors of production, in particular labor; (b) thus to ensure a fair standard of living for the agricultural community, in particular by increasing the individual earnings of persons engaged in agriculture; (c) to stabilize markets; (d) to assure the availability of supplies; (e) to

These objectives are closely aligned with various economic, social, and environmental goals recognized by the TFEU: promoting employment (Article 9), ensuring socio-economic cohesion (Articles 174–178), protecting the environment with a focus on sustainable development (Article 11), and safeguarding animal welfare (Article 13) and human health (Article 168).

Interpreted in this way, the CAP's legal framework appears to be well-aligned with the One Health model, as it effectively encompasses all its key pillars.[52]

Given that the CAP is a shared competence between the Union and its Member States, it is crucial to note that its governance at the European level is defined by specific hard law instruments. These include three key regulations that address the financing, management, and monitoring of the CAP, support for national CAP strategic plans, and the overall management of common agricultural markets.[53]

The balance between hard law interventions and vertical subsidiarity in the CAP enables its objectives to be achieved within the framework of Europe's "variable geometry".[54]

At the national level, there are 28 strategic plans in place (one for each Member State, except Belgium, which has separate plans for Flanders and Wallonia).[55]

Examining the various national strategic plans shows that the CAP's framework allows for flexible resource allocation. This flexibility enables each Member State

---

ensure that supplies reach consumers at reasonable prices.' 2. In working out the common agricultural policy, account shall be taken of: '(a) the particular nature of agricultural activity, which results from the social structure of agriculture and from structural and natural disparities between the various agricultural regions; (b) the need to effect the appropriate adjustments by degrees; (c) the fact that in the Member States agriculture constitutes a sector closely linked with the economy as a whole.'

[52] Nevertheless, there is notable scholarly critique. For instance, the necessity to clarify the CAP's role in human health, particularly from a nutritional standpoint, is discussed in Walls et al. (2016), pp. 12 ff.

[53] The reform in question comprises three regulations, listed as follows: Regulation (EU) 2021/2116 of the European Parliament and of the Council of 2 December 2021 on the financing, management, and monitoring of the common agricultural policy and repealing Regulation (EU) No 1306/2013; Regulation (EU) 2021/2115 of the European Parliament and of the Council of 2 December 2021 establishing rules on support for strategic plans to be drawn up by Member States under the common agricultural policy (CAP strategic plans) and financed by the European Agricultural Guarantee Fund (EAGF) and by the European Agricultural Fund for Rural Development (EAFRD), and repealing Regulations (EU) No 1305/2013 and (EU) No 1307/2013; Regulation (EU) 2021/2117 of the European Parliament and of the Council of 2 December 2021 amending Regulations (EU) No 1308/2013 establishing a common organization of the markets in agricultural products, (EU) No 1151/2012 on quality schemes for agricultural products and foodstuffs, (EU) No 251/2014 concerning the definition, description, presentation, labeling, and protection of geographical indications of aromatized wine products, and (EU) No 228/2013 laying down specific measures for agriculture in the outermost regions of the Union.

[54] In connection with this aspect, see Viaggi et al. (2013), pp. 99 ff.

[55] For a comprehensive overview of the key milestones within the various strategic plans, please refer to the document prepared by the European Commission (2022), titled *Common Agricultural Policy for 2023–2027: 28 CAP strategic plans at a glance*, available at the following link https://agriculture.ec.europa.eu/common-agricultural-policy/cap-overview/cap-2023-27_en.

to focus on areas of greatest concern while remaining within the scope of the CAP's overall objectives.[56]

For instance, consider the action plans of different Member States.

While Italy, Spain, and France primarily focus on enhancing the competitiveness and sustainability of their diverse agricultural sectors and rural areas, considering their regional differences, countries like Bulgaria and Romania concentrate on directing funding towards improving farmers' working conditions.

Considering the 28 CAP strategies, there are various connections with the One Health approach, which is inherently tied to the need for sustainable land use. Many national strategies emphasize reconnecting agriculture with a sustainability driver characterized by the interconnection between human, animal, and environmental health. This is achieved by channeling resources into organic farming and, more broadly, agroecology.

Some strategies also include provisions to improve animal welfare conditions, aligning with One Health objectives. Furthermore, significant references are made to the connection between human health and safety through the reduction of harmful pesticide use, which yields cascading benefits for soil, livestock, and human nutrition.

Additionally, references to enhancing the resilience of rural economies contribute to the holistic implementation of goals related to overall human well-being. In summary, many CAP strategies reflect the One Health approach, aiming for a coherent integration of various aspects of agricultural sustainability.[57]

---

[56] With reference to the 28 CAP strategies see Becker et al. (2022); Badouix et al. (2023).

[57] Several EU Member States have crafted Common Agricultural Policy (CAP) strategic plans that align closely with the principles of the One Health approach, which emphasizes the interconnectedness of human, animal, and environmental health. Here are a few notable examples. France's CAP plan prioritizes organic farming and agroecology, with ambitious targets for expanding organic cultivation. These initiatives support One Health by promoting agricultural practices that enhance soil health and minimize the use of harmful chemicals. The plan also incentivizes improvements in animal welfare, the reduction of antibiotic use, and the promotion of biodiversity, all of which contribute to healthier ecosystems and communities. Germany's strategy includes robust measures for biodiversity conservation, supporting sustainable agricultural practices that enhance habitat diversity. This approach is directly aligned with One Health goals, recognizing the link between biodiversity, human health, and animal welfare. The plan commits to significant reductions in pesticide use, reflecting a strong alignment with One Health principles by addressing the impact of pesticides on human health and biodiversity. Moreover, the Dutch CAP plan focuses on circular agriculture, aiming to reduce waste and close nutrient cycles. This strategy aligns with One Health by minimizing the environmental impact of agriculture and contributing to sustainable, resilient food systems. It places a significant emphasis on improving soil health through better management practices, erosion prevention, and the promotion of natural fertility—key factors in balancing environmental health with agricultural productivity. Italy's strategic plan strengthens the sustainability and resilience of rural areas, with targeted interventions to protect soil health and promote biodiversity. These measures support agricultural practices that minimize environmental impact and enhance the quality of life in rural communities. It also introduced policies aimed at reducing chemical use in agriculture, improving water management, and promoting safer, more sustainable food production.

Nevertheless, there are notable disconnections from this holistic model. Some strategies still seem overly fragmented, focusing predominantly on agricultural productivity without sufficiently addressing its health impacts or achieving the necessary balance. In other words, it is essential to remain vigilant regarding the persistent challenges associated with the environmental impact of intensive agriculture and the sustainable management of natural resources.[58]

If we consider that, in the 1960s, the CAP was initially established as an economic policy instrument aimed at ensuring the free movement of goods within the Community while maintaining public intervention in the agricultural sector—an area known for its inherent imbalances between supply and demand—we can now recognize it as an environmental policy tool. Over the years, it has evolved to implement a range of strategies focused on sustainability and environmental stewardship.

As discussed in the previous paragraphs, the current focus of environmental policy has shifted away from merely cataloging the challenges that hinder sustainable development. It now emphasizes the identification of a holistic approach that can effectively guarantee its achievement.

In this context, the new CAP can be considered a significant starting point for the concrete implementation of the One Health approach. While some inconsistencies still exist, there is undoubtedly a cohesive effort between European institutions and Member States in this direction.

5 Soil Health Through Market-Based Environmental Protection Instruments

Given the current state of European institutions' maturity in adopting the One Health approach and the significant gaps in its practical implementation, it is crucial to focus on a key issue: the role of legal instruments in environmental public policy aimed at promoting soil health.

These instruments are vital for aligning economic growth with ecosystem protection. Under the principles of European integration,[59] public authorities have their discretion constrained by sustainability requirements but remain central in shaping

[58] Here are examples of countries whose CAP Strategic Plans have been criticized for not fully aligning with the One Health principles: Poland's plan is focused on maintaining high agricultural productivity, particularly through intensive farming practices. This approach risks environmental sustainability, particularly soil health and biodiversity, which are key elements of the One Health framework; Hungary's plan has been noted for its emphasis on conventional farming methods, which may undermine environmental and animal health objectives. This focus could hinder the adoption of more sustainable agricultural practices; Romania's strategy has faced criticism for its emphasis on livestock production without sufficient attention to animal welfare and environmental impacts. This could result in practices that are not in line with the One Health approach, which seeks to balance agricultural productivity with the health of ecosystems, animals, and humans.

[59] For a general overview, see Wasmeier (2001), pp. 159 ff.; Wennerås (2008), pp. 163 ff.; Jans (2010), pp. 1533 ff.

the economic landscape.[60] Left to its own devices, the economic sector has shown a clear inability to self-regulate towards environmentally sustainable practices.[61] This situation underscores the need for new development paradigms and structural reforms, agreed upon at multiple levels.

This broader context is especially relevant to the specific issue at hand: the protection of soil as a cornerstone of the One Health approach.

This issue should be examined from two different perspectives. First, it is important to reassess the principle of European integration through a holistic approach like One Health.

According to Article 11 TFEU, the principle of integration requires that environmental protection needs be incorporated into the formulation and implementation of Union policies and activities to promote sustainable development. Today, it is evident that the foremost environmental protection priority is the interconnected health of humans, animals, and ecosystems, with the goal of achieving sustainable development where humanity plays a crucial role.

To better align European policies with this goal, placing greater emphasis on the broad concept of health within the integration principle could serve as a foundation for transitioning from fragmented policies to a cohesive One Health approach.

Secondly, the One Health perspective must be applied more effectively in practice. Here, legal instruments play a crucial role.

The main challenge is finding the right balance between hard and soft law ones.[62]

Hard law—linked to command and control mechanisms—is essential for setting common rules and minimum standards for the global agricultural market, while soft law helps guide the gradual integration of environmental considerations into trade practices. This approach is necessary given the diverse and multi-speed nature of the European political landscape.

Beyond the CAP eco-schemes, various market-based instruments[63] designed to protect the environment—some of which pertain to artificial markets—play a pivotal role in preserving soil health.

These market instruments can be categorized into two groups: those directly targeting the production sector and those also related to consumption aspects.

Starting with the former, Payments for Ecosystem Services (PES) offer financial incentives to farmers and landowners to maintain or enhance soil quality through

[60] The Commission has consistently advocated for this principle in various communications and within the context of environmental action programs. The integration of environmental considerations into the adoption of Union acts across other European policies is primarily facilitated through environmental impact assessments carried out by institutions during the legislative process. For an overview of the communications where the Commission has addressed the principle of integration, see Krämer (2004), pp. 34 ff. In particular, refer to the Commission Communication on the Integration of Environmental Aspects into European Legislation, COM (2004) 0130 final.

[61] Clarich (2007), pp. 219 ff.

[62] Among the various doctrinal contributions, the following are particularly highlighted: Holling and Meffe (1996), pp. 328 ff.; Harrington and Morgenstern (2004), pp. 13 ff.; Böcher (2012), pp. 14 ff.

[63] For the definition of these categories of instruments, please refer to Cafagno (2019), pp. 192 ff.

sustainable agricultural practices.[64] These incentives promote techniques such as crop rotation, maintaining vegetative cover, and minimizing the use of harmful chemicals.

Additionally, Agricultural Carbon Certificates[65] further contribute by allowing farmers to earn carbon credits for practices that increase soil carbon sequestration, such as conservation tillage or agroforestry.[66]

Regarding the encouragement of consumers to purchase sustainable agricultural products, attention should be given to certification systems and, in particular, to the Green Public Procurement (GPP) framework.

Sustainability certifications, including the EU organic label and other ecological certifications, encourage farmers to adopt practices that enhance soil health. These certifications reward products produced through methods that uphold environmental sustainability.

It is well established that public demand significantly influences the transition to an ecologically sustainable economy. In this regard, Green Public Procurement (GPP)[67] can play a crucial, albeit indirect, role in maintaining soil health. For example, in the management of public lands and green spaces, GPP policies can mandate

---

[64] The Payments for Ecosystem Services (PES) framework serves as a mechanism for the sustainable management of natural resources, emphasizing the preservation of ecosystem functionality, semi-natural areas, and the delivery of ecosystem services. These instruments are deemed essential for the effective implementation of the EU Biodiversity Strategy for 2030. Their significance is further underscored in the Communication from the Commission to the European Parliament, the Council, the European Economic and Social Committee and the Committee of the Regions, *Roadmap to a Resource-Efficient Europe*, COM (2011) 57. Among the most recent contributions on the subject, see Colognese and Granato (2024), pp. 57 ff.

[65] Agricultural Carbon Certificates are tradable credits awarded to farming operations for their efforts in sequestering carbon or reducing greenhouse gas emissions. These certificates, integral to carbon offset programs, represent a measurable amount of carbon dioxide (CO2) or its equivalent that has been either removed from the atmosphere or prevented from being released through specific agricultural practices. These instruments create a voluntary, non-mandatory market. However, those who choose to participate must adhere to the mandatory rules and regulations that govern the market to ensure the validity and integrity of transactions. Agricultural Carbon Certificates are regulated by various legal and regulatory frameworks. Key references include: EU Regulation (EU) 2018/842 that sets rules for greenhouse gas emission reductions by member states, impacting carbon credit systems; EU Regulation (UE) 2021/1119 that establishes climate neutrality goals and measures, relevant for agricultural carbon credits. Directive 2003/87/CE that creates the EU Emissions Trading System (EU ETS), influencing carbon credit mechanisms, including those in agriculture. Kyoto Protocol and Paris Agreement that outline emission reduction goals and impact the development of carbon credits. Furthermore, each country may have specific laws for agricultural carbon credits, aligning with broader EU and international policies.

[66] See Carsten et al. (2023), 117142.

[67] The term Green Public Procurement encompasses a range of legal tools aimed at incorporating environmental criteria into public procurement processes across all stages. The extensive efforts of the European Commission and the Court of Justice of the European Union have led to the development of directives that promote and expand sustainable public procurement practices. These include Directive 2014/24/EU, which governs procurement in general sectors; Directive 2014/25/EU, which addresses procurement in the water, energy, transport, and postal services sectors; and Directive 2014/23/EU, which focuses on concession contracts.

the use of environmentally friendly practices and materials in landscaping and construction. This might include selecting plant species that prevent soil erosion, implementing sustainable irrigation systems, and avoiding harmful pesticides and fertilizers—all of which contribute to protecting and improving soil quality.

Additionally, GPP can impact the procurement of agricultural products (such as through food procurement for school cafeterias). By incorporating specific environmental standards into bidding requirements or as criteria for awarding contracts, GPP policies can encourage the adoption of sustainable agricultural practices and, consequently, support soil health throughout the supply chain.

In conclusion, the potential of the legal instruments discussed is clearly aligned with the goal of implementing public policies that promote a One Health approach through a focus on soil health. Nonetheless, it is crucial to acknowledge that both the agricultural sector and consumers continue to experience substantial knowledge gaps in this area. Therefore, it is essential that public policies address this issue not only through legal measures but also through robust informational strategies to effectively advance and integrate the One Health approach.

6 Conclusions

The American political scientist David Easton famously stated that "public policy […] consists of a network of decisions and actions that allocate values." At its core, the concept of value allocation—a key aspect of public policy—raises an inevitable question: how should value be allocated? As we update this idea, it becomes clear that value allocation can no longer rely on isolated, sector-specific approaches.

Looking back to the beginning, we can now see that the interplay between risks and their far-reaching consequences demands a cross-cutting and integrated response. In this context, global public policy needs to break free from the narrow view that health pertains solely to humans and instead embrace the One Health concept.

The lens of the One Health approach offers a valuable tool for addressing the current polycrisis, ensuring that no one is left behind—not only humans but also animals and the environment. Based on undeniable scientific evidence, this paper underscores that from a One Health perspective, soil health must be recognized as a fundamental condition for life itself.

These directives clearly position public procurement as a critical element in advancing sustainability. Specifically, Recital 2 of Directive 2014/24/EU highlights that public procurement is a fundamental instrument for the Europe 2020 Strategy. It asserts that public procurement serves as a market-based tool essential for achieving smart, sustainable, and inclusive growth while ensuring the efficient use of public funds. For an in-depth analysis of the establishment of the institute at the European level, please refer to Depietri (2023), pp. 792 ff.

For insights into the role of public procurement in sustainability, please refer to Caranta and Cravero (2019), pp. 173 ff.; Mélon (2020), pp. 15 ff.; Valaguzza (2016, 2021); Halonen (2021), pp. 553 ff.

By examining key European strategies aimed at protecting soil health—both directly (e.g., Soil Strategy) and indirectly (e.g., From Farm to Fork Strategy and Biodiversity Strategy)—we observe a fragmented landscape. The back-and-forth between holistic and anthropocentric viewpoints suggests that while the EU is attempting to adopt this holistic approach, there remains a need for clearer direction and a more cohesive vision.

As these strategies seek to bridge principles and actions, it becomes crucial to highlight the One Health approach as a potential catalyst for advancing the concept of global health as a cornerstone of environmental protection. This, in turn, could reshape the principle of European integration, driving an ecological transition that is both economically competitive and ecologically comprehensive.

In this context, the agricultural sector, which is closely tied to soil health, emerges as a crucial area where the One Health approach could serve as a consistent standard for value allocation. The Common Agricultural Policy (CAP), as an environmental policy tool, has taken up this challenge, delivering both notable achievements and areas of inconsistency.

Amid this evolving landscape—marked by a tension between anthropocentrism and attempts to implement a holistic model—it is important to focus on practical considerations. Are there legal instruments suitable for implementing the One Health model? If so, which ones? While hard law instruments play a vital role in establishing minimum protection standards, attention should also be given to voluntary market tools. These can encourage the agricultural production and consumption sectors to adopt sustainable practices that prioritize soil health.

Today, public power plays a crucial role in driving the social, cultural, scientific, and institutional transition that is essential to achieving ecological sustainability. While the current situation may be complex and multifaceted, it can be distilled into a simple framework: problem—goal—tool.

The One Health approach sits at the heart of this framework, acting as a bridge between the problem and the goal, and between the goal and the tools needed to achieve it. On one hand, institutional awareness of the One Health approach facilitates an organic, interconnected analysis of needs. On the other, it helps identify the full range of cross-sectoral tools necessary to address those needs.

While a holistic approach is not a "cure-all" for every issue, its application allows for the systematization, evaluation, and resolution of problems in a coherent and integrated way.

Acknowledgments Research activity funded by a PNRR-MUR grant, Mission 4, Component 1 (Art. 6, Ministerial Decree No. 351/2022). CUP D92B22000500005

References

Atlas RM (2013) One Health: its origins and future. Curr Top Microbiol Immunol 365:1 ff.

Badouix M, Gorny H, Messinger I, Schuh B, Ade SA, Beck M, Bodart S, Van Bunnen P (2023) Comparative analysis of the CAP Strategic Plans and their effective contribution to the achievement of the EU objectives: research for AGRI Committee. European Parliament, Policy Department for Structural and Cohesion Policies. https://www.europarl.europa.eu/thinktank/en/document/IPOL_STU(2023)747255

Becker S, Grajewski R, Rehburg P (2022) Where does the CAP money go? Design and priorities of the draft CAP Strategic Plans 2023-2027. Thünen Working Paper, Münch, 191a

Beisheim M, Bernstein S, Biermann F (2022) Global governance. In: Biermann F, Hickmann T, Sénit CA (eds) The political impact of the sustainable development goals: transforming governance through global goals?. Cambridge University Press, pp 22 ff.

Bell J, Paula L, Dodd T, Németh S, Nanou C, Mega V, Campos P (2018) EU ambition to build the world's leading bioeconomy—uncertain times demand innovative and sustainable solutions. N Biotechnol 40(Part A):25 ff.

Berry EM, Dernini S, Burlingame B, Meybeck A, Conforti P (2015) Food security and sustainability: can one exist without the other?. Public Health Nutr 18:2293 ff.

Bevilacqua D, Chiti E (2024) Green deal. Come costruire una nuova Europa. il Mulino

Blaikie P (2016) The political economy of soil erosion in developing countries. Routledge

Blaikie P, Brookfield H (2015) Land degradation and society. Routledge

Böcher M (2012) A theoretical framework for explaining the choice of instruments in environmental policy. Forest Policy Econ 16(C):14 ff.

Brady NC, Weil RR (2007) The nature and properties of soils, 14th edn. Prentice-Hall, Upper Saddle River, New Jersey.

Bugge HC, Voigt C (2008) Sustainable development in international and national law. Europa Law Publishing

Cafagno M (2019) Strumenti di mercato a tutela dell'ambiente. In: Rossi G (ed) Diritto dell'Ambiente. Giappichelli, pp 192 ff.

Caranta R, Cravero C (2019) Sustainability and public procurement. In: La Chimia A, Trepte P (eds) Public procurement and aid effectiveness. Hart, Oxford, pp 173 ff.

Carsten P, Bartkowski B, Dönmez C, Don A, Mayer S, Steffens M, Weigl S, Wiesmeier M, Wolf A, Helming K (2023) Carbon farming: are soil carbon certificates a suitable tool for climate change mitigation? J Environ Manag 330:117142

Chomsky N, Pollin R (2020) The climate crisis and the global green new deal: the political economy of saving the planet. Verso Books

Clarich M (2007) La tutela dell'ambiente attraverso il mercato. Dir Pubbl, pp 219 ff.

Cleaveland S, Sharp J, Abela-Ridder B, Allan KJ, Buza J, Crump JA, Davis A, Del Rio VJ, de Glanville WA, Cassidy A (2018) Humans, other animals and 'One Health' in the early twenty-first century. In: Animals and the shaping of modern medicine. Medicine and biomedical sciences in modern history. Palgrave Macmillan, pp 193 ff.

Colognese C, Granato N (2024) Natural capital and ecosystem services-legal aspects. In: Antonelli G, Qin T, Ferroni MV, Erwin A (eds) Biodiversity laws, policies and science in Europe, the United States and China. Springer, Cham. pp 57 ff.

Cook RA, Karesh WB, Osofsky SA (2004) The Manhattan principles on 'One world One Health'. One World, One Health: building interdisciplinary bridges to health in a globalized world. https://www.oneworldonehealth.org/sept2004/owoh_sept04.html

Craddock S, Hinchliffe S (2015) One world, one health? Social science engagements with the one health agenda. Soc Sci Med, 129:1 ff.

Daszak P, Cunningham AA, Hyatt AD (2001) Anthropogenic environmental change and the emergence of infectious diseases in wildlife. Acta Trop, 78:103 ff.

De Leonardis F (2022) The ecosystem services provided by soil and the importance of its protection: the essential role of organic waste. RQDA 1:400 ff.

Depietri A (2023) Il Green Public Procurement tra riforme, mercato ed effettività. Munus 3:792 ff.

FAO (2020) State of knowledge of soil biodiversity - status, challenges and potentialities, FAO, ITPS, GSBI, SCBD, and EC. https://www.fao.org/policy-support/tools-and-publications/resources-details/en/c/1363310/

Faye JB, Braun YA (2022) Soil and human health: understanding agricultural and socio-environmental risk and resilience in the age of climate change. Health Place 77:102799

Forfora N, Azuaje I, Vivas KA, Vera RE, Brito A, Venditti R, Kelley S, Tu O, Woodley A, Gonzalez R (2024) Evaluating biomass sustainability: why below-ground carbon sequestration matters. J Clean Prod 439:140677

Fracchia F (2010) Lo sviluppo sostenibile. La voce flebile dell'altro tra protezione dell'ambiente e tutela della specie umana. Editoriale Scientifica

Glæsner N, Helming K, De Vries W (2014) Do current European policies prevent soil threats and support soil functions?. Sustainability 6(12):9538 ff.

Halonen KM (2021) Is public procurement fit for reaching sustainability goals? A law and economics approach to green public procurement. Maastricht J Eur Comp Law 28(4):553 ff.

Harrington W, Morgenstern R (2004) Economic instruments versus command and control: what's the best approach for solving environmental problems? Resources 2004:13 ff.

Helming K, Daedlow K, Hansjürgen B, Koellner T (2018) Assessment and governance of sustainable soil management. Sustainability 10(12):4432 ff.

Holling C, Meffe G (1996) Command and control and the pathology of natural resource management. Conserv Biol 10(2):328 ff.

Jans JH (2010) Stop the integration principle? Fordham Int Law J, pp 1533 ff

Kamenshchikova A, Wolffs PFG, Hoebe CJPA, Horstman K (2019) Anthropocentric framings of One Health: an analysis of international antimicrobial resistance policy documents. Crit Public Health 31:306 ff.

Kazwala RR et al (2017). One Health contributions towards more effective and equitable approaches to health in low- and middle-income countries. Philos Trans R Soc Lond B Biol Sci 372:1725 ff.

Keune H, Payyappallimana U, Morand S, Rüegg S (2022) One Health and biodiversity. In: Visseren Hamakers IJ, Kok MTJ (eds) Transforming biodiversity governance. Cambridge University Press, pp 93 ff.

Krämer L (2004) The genesis of EC environmental principles. In: Macrory R (ed) Principles of European environmental law. Groningen, pp 34 ff.

Lal R (2004) Soil carbon sequestration to mitigate climate change. Geoderma 123(1–2):1 ff.

Lal R, Bouma J, Brevik E, Dawson L, Field DJ, Glaser B, Hatano R, Hartemink AE, Kosaki T, Lascelles B, Monger C, Muggler C, Ndzana GM, Norra S, Pan X, Paradelo R, Reyes-Sanchez LB, Sandén T, Singh BR, Spiegel H, Yanai J, Zhang J (2021) Soils and sustainable development goals of the United Nations: an International Union of Soil Sciences perspective. Geoderma Reg 25:e00398

Mackenzie JS, Jeggo M (2019) The One Health approach-why is it so important? Trop Med Infect Dis 4:88 ff.

Mcbratney A, Field DJ, Koch A (2013) The dimensions of soil security. Geoderma 213:203 ff.

Mélon M (2020) More than a nudge? Arguments and tools for mandating green public procurement in the EU. Sustainability 12:15 ff.

Monteduro M (2015) From agroecology and law to agroecological law? Exploring integration between scientia ruris and scientia iuris. In: Monteduro M, Buongiorno P, Di Benedetto S, Isoni A (eds) Law and agroecology. Springer, Heidelberg, pp 57 ff.

Montgomery DR (2007) Soil erosion and agricultural sustainability. Proc Natl Acad Sci 104(33):13268 ff.

Morin E, Kern AB (1999) Homeland earth: a manifesto for the new millennium. Hampton Press

Musa TH, Ahmad T, Khan M, Haroon H, Wei P (2020) Global outbreak of COVID-19: a new challenge?. J Infect Dev Ctries 14(3):244 ff.

Nkonya E, Mirzabaev A, von Braun J (2016) Economics of land degradation and improvement – a global assessment for sustainable development. Springer, Cham

Novitz T, Pieraccini M (2020) Agenda 2030 and the sustainable development goals: "responsive, inclusive, participatory and representative decision-making"?. In: Pieraccini M, Novitz T (eds) Legal perspectives on sustainability. Bristol University Press, pp 39 ff.

Olofsson J, Börjesson P (2018) Residual biomass as resource–life-cycle environmental impact of wastes in circular resource systems. J Clean Prod 196:997 ff.
Panagos P, Standardi P, Borrelli P, Lugato E, Montanarella L, Bosello F (2018) Cost of agricultural productivity loss due to soil erosion in the European Union: from direct cost evaluation approaches to the use of macroeconomic models. Land Degrad Dev 29:383 ff.
Pe'Er G, Zinngrebe Y, Moreira F, Sirami C, Schindler S, Müller R, Bontzorlos V, Clough D, Bezák P, Bonn A, Hansjürgens B (2019) A greener path for the EU Common Agricultural Policy. Science 365(6452):449 ff.
Pe'Er G, Finn JA, Díaz M, Birkenstock M, Lakner S, Röder N, Guyomard H (2022) How can the European Common Agricultural Policy help halt biodiversity loss? Recommendations by over 300 experts. Conserv Lett 15(6):e12901
Pettifor A (2019) The case for the green new deal. Verso Books
Philp J (2018) The bioeconomy, the challenge of the century for policy makers. N Biotechnol 40(Part A):11 ff.
Pitto S (2023) Cambiamento climatico e sicurezza alimentare: dall'approccio One Health ai modelli olistici del Global South. BioLaw J - Rivista Di BioDiritto 2:315 ff.
Prentice H, Ohlson K (2014) The soil will save us: how scientists, farmers, and foodies are healing the soil to save the planet. Rodale Books
Rabinowitz P, Conti L (2013) Links among human health, animal health, and ecosystem health. Annu Rev Public Health 34(1):189 ff.
Rajamani L, Peel J (2021) The Oxford handbook of international environmental law. Oxford University Press
Ray DK, West PC, Clark M, Gerber JS, Prishechepov AV, Chatterje S (2019) Climate change has likely already affected global food production. PLoS One 14(5)
Röder N, Krämer C, Grajewski R, Lakner S, Matthews A (2024) What is the environmental potential of the post-2022 common agricultural policy? Land Use Policy 144:107219
Roeben V (2020) The law of sustainable development: general principles. Intersentia
Sachs JD (2014) The age of sustainable development. Columbia University Press
Sands P, Peel J, Fabra A (2018) Principles of international environmental law. Cambridge University Press
Sanz-Hernández A, Esteban E, Garrido P (2019) Transition to a bioeconomy: perspectives from social sciences. J Clean Prod 224:107 ff.
Schebesta H, Chandel JL (2020) Game-changing potential of the EU's Farm to Fork Strategy. Nat Food, pp 586 ff.
Schrijver N, Weiss F (2004) International law and sustainable development: principles and practice. Martinus Nijhoff
Scotti E (2023) One Health: per un'integrazione tra salute umana e ambientale. In: Aperio Bella F (ed) One Health: la tutela della salute oltre i confini nazionali e disciplinari. Editoriale Scientifica, pp 45 ff.
Sweeney JM, Crook ZP, Deeb-Soss N, Tu B, Dear JD, Mazet JAK (2018) Clinical one health: a novel healthcare solution for underserved communities. One Health 6:34 ff.
Tilman D, Clark M (2014) Global diets link environmental sustainability and human health. Nature 515:518 ff.
Valaguzza S (2016) Sustainable development in public contracts: an example of strategic regulation. Editoriale Scientifica
Valaguzza S (2021) Governing by contract procuring for value. Creating value through public contracts, Editoriale Scientifica
Van Potter R (1971) Bioethics: bridge to the future. Prentice-Hall
Veerman C, Pinto Correia T, Bastioli C, Biro B, Bouma J, Cienciala E, Emmett B, Frison EA, Grand A, Hristov L, Kriaučiūnienė Z, Pogrzeba M, Soussana JF, Vela CO, Wittkowski R (2020) Caring for soil is caring for life. Ensure 75% of soils are healthy by 2030 for food, people, nature and climate: report of the Mission board for Soil health and food, Publications Office of the EU. https://op.europa.eu/en/publication-detail/-/publication/4ebd2586-fc85-11ea-b44f-01aa75ed71a1/

Venturi F (2021) The farm to fork strategy. A comprehensive but cautious approach to "multidimensional" food sustainability. RQDA 1:70 ff.

Viaggi D, Gomez y Paloma S, Mishra A, Raggi M (2013) The role of the EU Common Agricultural Policy: assessing multiple effects in alternative policy scenarios. Land Use Policy 31:99 ff.

Vivien FD, Nieddu M, Befort N, Debref R, Giampietro M (2019) The hijacking of the bioeconomy. Ecol Econ 159:189 ff.

Wallace RG, Bergmann L, Kock R, Gilbert M, Hogerwerf L, Wallace R, Holmberg M (2015) The dawn of structural One Health: a new science tracking disease emergence along circuits of capital. Soc Sci Med 129:68 ff.

Walls HL, Cornelsen L, Lock K, Smith RD (2016) How much priority is given to nutrition and health in the EU Common Agricultural Policy?. Food Policy 59:12 ff.

Wasmeier M (2001) The integration of environmental protection as a general rule for interpreting community law. Common Mark Law Rev, pp 159 ff.

Wennerås M (2008) Towards an ever greener Union? Competence in the field of the environment and beyond. Common Mark Law Rev, pp 163 ff.

Climate Change and a Comprehensive Reparation of Damaged Interests: A Cultural Issue

Anna Malomo

Abstract The need to adequately compensate for damage to interests, especially of a non-pecuniary nature, caused by reckless behaviour towards climate change is now unavoidable. This is known as climate change damages. The consequences will be felt in time and space by younger and future generations. It is right to speak of damage that affect the existential legal sphere of individuals who are *a priori* determinable, for whom it is unthinkable to continue to invoke the unpredictability of the events that caused them. A punitive compensation may (also) prove to be the appropriate remedy, since it is at the same time: dissuasive, to prevent the continuation of prejudicial behaviour from the point of view of intergenerational solidarity; exemplary, to create an appropriate cultural sensitivity to the issue; and functional, in the immediate future, to fully compensate the damage suffered.

1 Introduction

Aware of the plurality of functions that civil liability fulfils,[1] it is necessary to identify appropriate remedies for damage, especially non-pecuniary, caused as a result of the unwise political choices made by States around the world. There is also the wrongful behaviour of other actors (companies, multinationals, etc.) responsible for

[1] Perlingieri (P 2008), p. 254; Perlingieri (P 2011), pp. 115 ff. and especially 117.

A. Malomo (✉)
Department of Law, Università degli Studi di Salerno, Fisciano, Italy
e-mail: amalomo@unisa.it

G. Antonelli et al. (eds.), *Planetary Health - Laws, Policies and Science on the 'One Health' Approach*, https://doi.org/10.1007/978-3-031-90621-3_3

the increase in greenhouse gas emissions and the disasters they cause:[2] climate change damage.[3]

2 Climate Change and Damage

The damage being caused by climate change is of considerable magnitude for the whole of humanity, both present and future.

It is taking place at a relentless pace in violation of European and international constitutional principles.[4] Recalled by the Paris Agreement (COP 21) of 2015, in implementing the Cancun Agreements (COP 16) of 2010—which set the objective of limiting the average increase in the Earth's temperature—adequate protection of the environment and (climate) global primacy are necessary[5] and certainly not "on a par with economic and social interests". It has been constantly part of the agenda, starting from the United Nations Framework Convention on Climate Change (UNFCCC) of 1992 up to the Dubai Climate Conference (COP 28) of December 2023.[6] The latter ended with a commitment by the Paris signatory States to "abandon fossil fuels in their energy systems" by 2035. This was reaffirmed at the G7 Italy Ministerial Meeting on Climate, Energy and Environment.[7] This is an important commitment that has been required for more than a decade by the Italian Constitutional Court, which has qualified it as a public interest, or rather a general interest,[8] according to the best doctrine that identifies the public interest with "the realisation and implementation of inviolable human rights".[9] This is therefore a sig-

[2] The earth's temperature is estimated to rise by 2 °C despite the proclaimed attempt to contain the increase to within 1.5 °C of pre-industrial levels. See: Lipari (2024), pp. 211 f.; Zarro (2022), pp. 99 ff.

[3] For further details, see: Carducci (2021); Puleio (2021), p. 479.

[4] Fundamental under *ex* Art. 2 of the Italian Constitution.

[5] See Montini (2020), pp. 507 and 512.

[6] See the Resolution 76/300 of the 76th Session—UN General Assembly, 28 July 2022, §§ 2–4, which states that the impact of climate change, together with pollution, consequent biodiversity loss and environmental damage, has negative implications, both direct and indirect, for the effective enjoyment of all human rights; so urgent is the call for States, international organisations and businesses to implement more intensively the inescapable actions to ensure a clean, healthy and sustainable environment for all, in implementation of the multilateral agreements already in place and longstanding. For further details, see Pauciulo (2022), pp. 1118 ff.

[7] Turin, 28–30 April 2024, https://www.g7italy.it/en/about-g7, where the intention of decarbonisation within the next decade was confirmed. Critical of the incisiveness of the actions announced, see Quaranta (And, 2024), pp. 261 ff.; along with Tozzi (2023).

[8] According to Italian constitutional jurisprudence, it is in the public interest to "eliminate the dependence on fossil fuels" (Italian *Constitutional Court 1 April 2010, no. 124; Italian Constitutional Court 23 December 2019, no. 286; Italian Constitutional Court 13 November 2020, no. 237; Italian Constitutional Court 17 November 2021, no. 46:* https://cortecostituzionale.it).

[9] Perlingieri (P 1989a), pp. 264 f.; Perlingieri (P 2003a), pp. 58 f.; and, accordingly, Federico (1999), pp. 34 f. and 39.

nificant commitment, while implementing the principles of reasonableness[10] and solidarity.[11] In the face of offences that have already been committed, inertia[12] and even less impunity[13] cannot be justified.

The normative value of the values of the individual, of their dignity and of solidarity, including intergenerational solidarity,[14] together with the right *to* life and respect *for* private and family life, in accordance with the European Convention on Human Rights,[15] constitute for the interpreter who has been called upon to defend them[16] the "grounds for decision" and the "normative path" to be followed in the defence of the claims of individuals who, by virtue of the interests worthy of protection to be asserted, are legitimated to act (art. 24 of the Italian Constitution).

3 Damage to the Individual and Their Fundamental Rights

The change of perspective on climate issues is to be welcomed, as the concern for fundamental rights focuses on the point at which the relevant claims for protection must be centred: the individual and their fundamental rights, not the climate.[17] This implies that those directly responsible must be held accountable for the damage concretely caused to existential situations[18] (Art. 2, 3, 9, 24 and 32 of the Italian Constitution).[19]

[10] *Ex* Art. 32 of The Vienna Convention on the Law of Treaties of 1969.

[11] *Decision-/CMA.5 Outcome of the first global stocktake*, 13 December 2023, https://unfccc.int. See, Carducci (2023).

[12] In this regard, the considerations made by Mastrojeni and Pasini (2020), p. 149, are provocative. "Sed fugit interea fugit irreparabile tempus!" (Virgilio, III, p. 284).

[13] See Visintini (2009), pp. 42 f.

[14] Art. 2, 3, 9 and 32 of the Italian Constitution. Salanitro (2024), p. 251, considers essential "the reference, contained in the constitutional reform, to the protection of the interests of future generations", to allow the "interpreter" to identify the "subjects legitimated to act in the interest of those who [...] not yet conceived, [...] are exposed to the risks of climate change".

[15] Art. 2 and 8 of the European Convention of Human Right (ECtHR). The respect for private and family life coincides with the enjoyment of the amenity of one's home and generally one's "living space" (formerly ECtHR 21 February 1990, Application no. 9310/81, *Powell & Rauner v. UK*, § 40, https://echr.coe.int).

[16] See instead Meli (2023), pp. 294 f.

[17] This change is clearly of anthropogenic origin, as confirmed by the 2014 report of the Intergovernmental Panel on Climate Change (IPCC 2014). Anthropogenic activity is particularly evident in the diffusion of very fine gases and dust into the atmosphere, see Meli (2023), pp. 294 f.

[18] Pope Francis (2015) highlighted the need to protect people first, along with the environment and climate.

[19] In general, also because of the impossibility of living in a climate system that is reasonably adequate to guarantee a responsive human quality of life. In this context, the considerations of the US District Court for the District of Oregon, in the person of Judge Aiken, in the Juliana case (see Juliana v. USA, 217 F Supp 3d 1224 (D Ore 2016), pp. 18 f.) are interesting. She discusses an implicit constitutional right to live in a climate system capable of sustaining human life as func-

It is interesting to note the various decisions taken by judicial bodies in other countries,[20] and subsequently by others, condemning the States concerned, and sometimes certain multinationals. This confirms the significant increase in the demand for justice in the field of climate change.[21]

In Italy, in 2021, two appeals were brought before the European Court of Human Rights (ECHR). They were for the violation of the right to respect for private and family life as well as the principle of non-discrimination (Art. 2, 8 and 21 ECHR) in the knowledge that the climate issue "knows no borders"[22] due to the damage it causes, with huge repercussions on people's living conditions and health.[23] At the Court of Rome, a number of organisations fighting for the environment and very young people have brought a case called "Universal Judgement"[24] against the Italian State. As stated in the summons,[25] the State is not very receptive to the fight against greenhouse gas emissions. It is therefore called to account to the community for the so-called climate commitment and the psycho-physical well-being and health of all humanity, pursuant to articles 2043 and 2051 of the Italian Civil Code, in accordance with art. 2 and 97 of the Italian Constitution.[26] This is because it is the fundamental right of every individual to an optimal quality of life that is functional to their wellbeing that is being violated. This has been well outlined since the 2014 Conference of the "Italian Society of Civil Law Scholars (SISDiC)" on the subject

tional to the realisation of a free and ordered society. She bases her arguments on a precedent (Obergefell v. Hodges) which outlined an evolutionary interpretation of fundamental rights, so that notions of justice and injustice, and of needs worthy of protection, are responded to differently over time, as society changes. In this respect, Ghinelli (2021), pp. 1289 f., "finds in this passage a certain similarity with the reading of art. 2 [of the Italian Constitution] as an open-ended clause, the matrix of the rights of the "new generation"". Perceptive on this point, see Perlingieri (P 1972), *passim* and especially pp. 16 ff., 133 f., 139, 175 ff. and 178 f.

[20] There are more than 2000 pending cases, as recorded in two databases set up at two main international research centres: the Sabin Centre for Climate Change Law at Columbia University of New York (in https://climatecasechart.com) and the Grantham Research Institute on Climate Law and the Environment at the London School of Economics (in https://climate-laws.org).

[21] See *Mullner v. Austria* (Application no. 18859/21), *Greenpeace Nordic and Others v. Norway* (Application no. 34068/21), *Soubeste and four other applications v. Austria and 11 Other States* (Applications nos. 31925/22, 31932/22, 31938/22, 31943/22 and 31947/22), *Engels c. Germany* (Application no. 46906/22): https://echr.coe.int/Documents/FSClimatechangeENG.pdf. For further details, see Osti (2023), pp. 239 f.

[22] This was highlighted by the defence of the young woman from Matera: see Moliterni (2023).

[23] Applications Uricchio v. Italia and 31 Other States (no. 14615/21) and De Conto v. Italia and 32 Other States (no. 14620/21), https://echr.coe.int/Documents/FSClimatechangeENG.pdf.

[24] Case *Universal Judgement*, https://giudiziouniversale.eu, which has recently come to a halt: see footnote no. 31 below. On the subject, see Puleio (2021), pp. 491 f.

[25] See Case "A SUD and others Vs Italian State", (summons), https://giudiziouniversale.eu.

[26] Of the dangerous situation resulting from its inactivity in the fight against the climate emergency (it is required to proceed abruptly to reduce greenhouse gas emissions by 92% by 2030 compared to 1990 levels, pursuant to art. 2058 of the Italian Civil Code and in accordance with the commitment made at COP 28).

of "Wellbeing and the Rules of Civil Relations. Development beyond the crisis".[27] The Italian State is responsible for omission and, before that, for disinformation, in defiance of the constitutional principle of good administration to which its actions should conform.

It is also worth noting that Article 2059 of the Italian Civil Code has not been evoked. This is despite the fact that the unanimous Italian jurisprudence, now 20 years old, constitutional and legitimate, supported by a majority doctrine,[28] continues to proclaim the bipolarity of the civil liability system. It also typifies the so-called non-pecuniary damage, by limiting it to article 2059 of the Italian Civil Code. It deliberately disregards the exhaustiveness of Article 2043 of the Italian Civil Code, as a general open clause, by implementing the principle of *neminem laedere*, according to a reading of the law in conformity with the Italian Constitution. For some time now, the best doctrine has argued:[29] "responsibility does not presuppose the typicality of the tort", which is "a prehistory", and personality rights, necessarily "an open and constantly evolving number", confirm that "this atypicality does not prevent responsibility".[30] Therefore, the application of Art. 2043 of the Italian Civil Code, declined in accordance with Art. 2 of the Italian Constitution, is fully compliant to ensure the adoption of the most appropriate remedy in the face of the damage of any interest, patrimonial or not. This is even more so when one or more fundamental rights of the individual are violated by an inept or at least ineffective fight against climate change.

---

[27] In view of the "damage to non-patrimonial assets": see Perlingieri (P 2015), pp. 509 et seq.; see also Perlingieri (P 2016), p. 549.

[28] See Supreme Court of Cassation of Italy 31 May 2003, nos. 8827 and 8828, Il Foro Italiano, I, col. 2272 ff. (2003) [as well as Supreme Court of Cassation of Italy 19 August 2003, no. 12124, Il Foro Italiano, I, col. 434 ff. (2004) with a commentary by Costanza (2004)], which, with the approval of the Italian Constitutional Court [Italian Constitutional Court of Italy 11 July 2003, no. 233, Rassegna di Diritto Civile, pp. 770 ff. (2003), with a commentary by Perlingieri (P 2003b), pp. 775 ff.], erroneously arrives at the "implementation of constitutional principles and values" by means of "a primary source norm" (art. 2059 of the Italian Civil Code), but even more so to ""invest" the same norm with two contextual *rationes*: one, to protect the subjective damage suffered, and the other, to forcibly safeguard constitutionally protected fundamental human rights", also at European and international levels, "with contextual impoverishment of the centrality of art. 2043" of the Italian Civil Code. This "constitutional interpretation" of 2003—confirmed in 2008 by the Court of Cassation (Italy)—United Sections, 11 November 2008, no. 26972 [Rassegna di Diritto Civile, pp. 499 ff. (2009), with a critical commentary by Perlingieri (P 2009), pp. 520 ff.]—is based on the recognised primacy of the position of the Italian Constitution in the current legal system and on the scope of Art. 2 of the same Italian Constitution, which, in recognising and guaranteeing the inviolable rights of the individual, requires that non-pecuniary damage be understood "in its broader meaning of damage determined by the violation of interests inherent to the individual, not connoted by economic importance". See Navarretta (2009), pp. 84 f., along with Meli (2023), p. 302; commenting on the judgement of inadmissibility in question, see Foffa (2024).

[29] Perlingieri (P 2009), p. 529; Perlingieri (P 2020a), pp. 306 f. and footnote no. 6, and 341 ff.; Perlingieri (G 2010), pp. 399 ff.; Perlingieri (G 2015), p. 37 and footnote no. 75.

[30] Precursorily, Perlingieri (P 1989b), p. 138.

In February 2024, the petition for justice filed with the Court of Rome found its first halt for "supposed" incompetence,[31] but this will not stop the defence of fundamental rights undertaken.

If, on the one hand, it seems correct to speak of long-term damage and with an impact on existential legal situations, the potential owners of which are individuals who cannot be determined *a priori*. On the other hand, it cannot be denied that they are configurable.[32] All of these hypotheses call for justice, and most of them have so far been brought forward through collective actions, by organisations representing general interests and, more recently, by young or very young people.

The considerations of the most sensitive doctrine on the legal relevance of the subjective situation regardless of the current identification of its holder[33] are illuminating. There may be hypotheses in which there is already a relationship between certain centres of interest, but the holder of one or both does not yet exist or, if they do, will be affected at some future time.[34] The survival of human life itself, when it is increasingly compromised as a result of climate change, constitutes an existential subjective situation, regardless of whether or not its holder is still alive.[35] Such a damaged interest, of a non-pecuniary nature, cannot but be remedied.[36] The subjective existential situation thus configured exists, but it requires, if the individual responsible is identified, full and proportionate reparation. It should also be dissuasive and punitive to discourage such behaviour.[37] If, on the other hand, the individual is not yet identified, a firm condemnation to take proactive measures to mitigate climate change, in application of the constitutional principles of solidarity and subsidiarity (art. 2 and 118, par. 4, of the Italian Constitution).

On the other hand, a multiplicity of causal contributions, proven by innumerable scientific reports, allows both the individual-victim, present or not yet born, and a

[31] Court of Roma (Italy), 26 February 2024, https://ambientediritto.it.

[32] Lipari (2024), p. 224, points out that "in the changed ecological-institutional scenario, law, including private law, is called upon to face new possibilities of damage and danger, previously unthinkable or unknown, which correspond to protection requirements worthy of protection" by the Italian legal system.

[33] Perlingieri (P 1975, reprint 2019), pp. 37 ff. Finally, Tedoldi (2024) also points in this direction, asserting the need for "remedies – both substantive and procedural, without any distinction that is now obsolete" to consider the "growing complexity" that characterises our reality, given by "changes, wars, arms races, nuclear threats, pandemics, famines, floods, earthquakes, recurrent crises [. …] inequalities, job insecurity, abnormal asymmetries of information and technological, military and economic-financial power".

[34] Perlingieri (P 2020b), p. 271, specifies: "There are hypotheses in which there is a relationship between determined and legally relevant centres of interest that do not yet have a holder or have lost one (right to life of the conceived, right to honour of the deceased, donation to the unborn or unconceived, and others): in these the actuality of the subject is not essential to the existence of the situation. This means that a legally relevant relationship can exist between two or more centres of interest and that it cannot be defined as a relationship between subjects". Even earlier, see Perlingieri (P 1968), p. 83, footnote no. 197; Perlingieri (P 1979), p. 266; similarly, even if life has been lost, full compensation must be made for the maximum damage suffered [Malomo (2015), pp. 404 ff.] For further details, see also Perlingieri (P 1975, reprint 2019), pp. 37 ff.

[35] For further details, see Perlingieri (G 2018), p. 726.

[36] See Lipari (2024), p. 218.

[37] On the punitive function, see Malomo (2017), *passim*.

professional association[38] that protects existential situations, to hold the perpetrators directly responsible and in proportion to their contribution for the damage thus caused and/or perished.[39]

The unpredictability of the resulting catastrophic events can no longer be invoked. More than one pronouncement, even in Italy, excludes such unpredictability in the case of serious damage caused by sudden rainfall of considerable intensity, with the responsibility of the public administration on duty, as custodian, therefore required to maintain the territory under its jurisdiction.[40]

Similarly, intense heat waves, which often lead even those who work outdoors to seek legal redress, can no longer be defined as unpredictable. The issue is particularly relevant in terms of the fundamental rights at stake (loss of life and outbreaks of disease).[41] In this context, two decisions of the Court of Palermo (Italy)—Labour Section—of August 2022,[42] triggered by stories of discomfort experienced by two *riders*, prove interesting.

4 Possible Remedies

In view of the violations of such fundamental rights caused by climate change, it therefore seems reasonable that a functional remedy should be guaranteed to compensate for any damage suffered. It should include an adequate personalisation of the damage to be guaranteed both to the individual victims as well as their closest

[38] ECHR-GC, 9 April 2024, Application no. 53600/20, Affaire Verein Klimaseniorinnen Schweiz et autres v. Suisse, https://hudoc.echr.coe.int, confirms the legal status of professional associations.

[39] In order to challenge the environmental policies adopted by the European Union in this respect, it is possible to bring an action under Art. 263 of the FEU Treaty, as in the *People's Climate* case (*Carvalho case*: Court of Justice of the European Union, 25 March 2021, C-565/19, Armando Ferrão Carvalho and Others v. European Parliament and Council of the European Union, https://curia.europa.eu) and, in the case of contested infringements of fundamental rights, claim the liability of the Union itself under Art. 340, § 2, of the FEU Treaty.

[40] Art. 2043 and 2051 of the Italian Civil Code. See Supreme Court of Cassation (Italy) 24 March 2016, no. 5877, Danno e Responsabilità, pp. 839 ff. and especially 840 (2016). In this direction—but for cases that occurred in years in which the causal link with climate change was not yet configurable—see: Supreme Court of Cassation (Italy) 20 February 2006, no. 3651, Giurisprudenza Italiana, pp. 1403 ff. (2007); Supreme Court of Cassation (Italy) 8 May 2013, no. 10898, Diritto & Giustizia, pp. 554 ff. (2013). Also see, Supreme Court of Cassation (Italy) 28 July 2017, no. 18856, Diritto & Giustizia, f. 128, p. 6 (2017) (with a favourable commentary by Savoia (2017)), states that climate changes, together with the "well-known hydrogeological disruptions that characterise our country", no longer allow us to consider as unforeseeable events atmospheric phenomena (such as the so-called "water bombs") "increasingly frequent".

[41] So much so that INAIL was called upon to set up the Worklimate project, https://www.worklimate.it, and more recently the National Labour Inspectorate issued a note on limiting the risks of so-called heat stress, based on two court rulings in Palermo (see the following note).

[42] Court of Palermo (Italy)—Labour Section 18 August 2022, https://www.wikilabour.it/wp-content/uploads/2022/09/20220818_Trib-Palermo.pdf; Court of Palermo (Italy)—Labour section 3 August 2022, https://www.wikilabour.it/wp content/uploads/2022/09/20220803_Trib Palermo.pdf. See Del Frate (2023), pp. 463 ff.

relatives, who are affected in various ways by severe physical, psychological and relational suffering. This has been clearly stated for over a decade by the Third Chamber of the Italian Court of Cassation,[43] based on the intensity and duration of the catastrophic event suffered,[44] the effects caused to psycho-physical integrity (defined as *pathoclimate*), to the environment as well as including the loss of life itself.

Furthermore, Italian jurisprudence and the ECtHR have long implemented a wide-ranging protection of human health. This includes the potential risk of contracting diseases and the consequential compensability of the personal, family and relational distress experienced as a result of the periodic medical check-ups to monitor a physical condition that has unexpectedly deteriorated. Examples include the environmental disasters of Seveso with the toxic cloud,[45] Taranto with the Ilva[46] and Campania with the waste crisis.[47] The latter was finally condemned in October 2023 for failing to protect the health of the people involved.[48]

The adoption of remedies that respond to the need to protect the individuals concerned seems to be the optimal goal today. This can be achieved through the aforementioned "regulatory path". It has long been followed and is proving to be the only viable one[49] to encourage everyone, responsibly and consciously, at every level, with the appropriate cultural sensitivity, to change course and move towards prevention for the protection of people, present and future, and the sustainability of the environment, "in a personalistic and solidaristic key".[50]

[43] See: Supreme Court of Cassation (Italy) 20 November 2012, no. 20292, Danno e Responsabilità, pp. 129 ff. (2013); Supreme Court of Cassation (Italy) 22 April 2024, no. 10787, https://quotidianogiuridico.it.

[44] Heat waves, droughts, fires, floods, rising sea level, lakes or other watercourses, salination of soils, drinking water shortages, malnutrition.

[45] See Supreme Court of Cassation-United sections (Italy) 21 February 2002, no. 2515, Danno e Responsabilità, pp. 499 ff. (2002).

[46] ECHR 24 January 2019, Application no. 54414/13 e 54264/15, Cordella and Others v. Italia, https://hudoc.echr.coe.int, as well as ECHR 5 May 2022, Application no. 4642, Ardimento and c. v. Italia, https://hudoc.echr.coe.int. See Salanitro (2015), pp. 585 ff.; Meli (2020), pp. 477 ff.

[47] ECtHR 10 January 2012, Application no. 30765/08, Di Sarno and Others v. Italia, Il Foro Italiano, IV, col. 205 ff. (2013).

[48] ECtHR 19 October 2023, Application no. 35648/10, Locascia and Others v. Italia, https://quotidianogiuridico.it (26 October 2023).

[49] Valaguzza (2021), pp. 293 ff. argues otherwise. See *retro* §§ 1, 2 and 3.

[50] Sustainability "means [...] considering the impact on the environment and, conversely, the impact on the environment is a condition for the "sustainability" of any activity": "such sustainable social and economic progress [always] implies respect for pre-existing cultural values and identities" [Perlingieri (P 2020a), pp. 255 f.] and sustainability is its "social expression" [Caterini (2018), p. 149].

References

Carducci M (2021) Cambiamento climatico (diritto costituzionale). Digesto delle discipline pubblicistiche, VIII News update, Torino, Utet, pp 51–73
Carducci M (2023) Le novità della COP 28 tra uso delle parole e Costituzione, 17 December. https://lacostituzione.info
Caterini E (2018) Sostenibilità e ordinamento civile. Per una riproposizione della questione sociale. Edizioni Scientifiche Italiane, Napoli
Costanza M (2004) Ancora sul danno esistenziale. Il Foro Italiano, I, col. 434–436
Del Frate M (2023) Est modus in rebus. La gestione dello stress termico da calore nel settore del food delivery. Diritto delle Relazioni Industriali 2:463–469
Federico A (1999) Autonomia negoziale e discrezionalità amministrativa. Gli "accordi" tra privati e pubbliche amministrazioni. Edizioni Scientifiche Italiane, Napoli
Foffa R (2024) Cambiamento climatico? Inammissibili le richieste di risarcimento allo Stato, 3 Avril. https://quotidianogiuridico.it
Ghinelli G (2021) Le condizioni dell'azione nel contenzioso climatico: c'è un giudice per il clima? Rivista Trimestrale di Diritto e Procedura Civile, pp 1273–1297
IPCC (2014) Climate Change 2014 Synthesis Report, summary for policymakers. https://ipcc.ch
Lipari N (2024) Premesse per un diritto civile dell'ambiente. Rivista di Diritto Civile 2:209–228
Malomo A (2015) Perdita della vita e riparazione integrale. Diritto delle Successioni e della Famiglia, pp 383–417
Malomo A (2017) Responsabilità civile e funzione punitiva. Edizioni Scientifiche Italiane, Napoli
Mastrojeni G, Pasini A (2020) Effetto serra, effetto guerra. Il clima impazzito, le ondate migratorie, i conflitti. Il riscaldamento globale, i ricchi, i poveri. Chiarelettere, Milano
Meli M (2020) The environment, health and employment: Ilva never-ending story. Italian Law J, pp 477-504
Meli M (2023) Inquinamento atmosferico. Inadempimento agli obblighi imposti dalle direttive europee e strumenti di tutela. Nuova Giurisprudenza Civile Commentata I:294–308
Moliterni F (2023) 2 lucane e 2 bellunesi ad un passo dall'impresa di inchiodare l'Italia, 23 August. https://lecronachelucane.it
Montini M (2020) Verso una giustizia climatica basata sulla tutela dei diritti umani. Ordine Internazionale e Diritti Umani, pp 506–537
Navarretta E (2009) Danni non patrimoniali: il compimento della Drittwirkung e il declino delle antinomie. Nuova Giurisprudenza Civile Commentata II:81–89
Osti A (2023) A qualcuno (non) piace caldo. Il caso KlimaSeniorinnen c. Svizzera avanti alla Corte europea dei diritti dell'uomo (per non tacer degli altri). BioLaw J – Rivista di BioDiritto, vol 2, pp 237–251
Pauciulo D (2022) Il diritto umano a un ambiente salubre nella risoluzione 76/300 dell'Assemblea generale delle Nazioni Unite. Rivista di Diritto Internazionale, pp 1118–1125
Perlingieri G (2010) Sul giurista che come "il vento non sa leggere". Rassegna di Diritto Civile, pp 385–413
Perlingieri G (2015) Profili applicativi della ragionevolezza nel diritto civile. Edizioni Scientifiche Italiane, Napoli
Perlingieri G (2018) Ragionevolezza e bilanciamento nell'interpretazione recente della Corte costituzionale. Rivista di Diritto Civile, pp 716–753
Perlingieri P (1968) Remissione del debito e rinunzia al credito. Jovene, Napoli
Perlingieri P (1972) La personalità umana nell'ordinamento giuridico. Edizioni Scientifiche Italiane, Napoli-Camerino
Perlingieri P (1975, reprint 2019) Dei modi di estinzione delle obbligazioni diversi dall'adempimento. Edizioni Scientifiche Italiane, Napoli
Perlingieri P (1979) Profili istituzionali del diritto civile, 2th edn. Edizioni Scientifiche Italiane, Napoli

Perlingieri P (1989a) Il diritto agrario tra pubblico e privato (1985). In: Perlingieri P (ed) Scuole tendenze e metodi. Problemi del diritto civile. Edizioni Scientifiche Italiane, Napoli, pp 257–271
Perlingieri P (1989b) Spunti in tema di tutela dell'ambiente. Legalità e Giustizia:136–139
Perlingieri P (2003a) L'incidenza dell'interesse pubblico sulla negoziazione privata (1986). In: Perlingieri P (ed) Il diritto dei contratti fra persona e mercato. Problemi del diritto civile. Edizioni Scientifiche Italiane, Napoli, pp 53–69
Perlingieri P (2003b) L'art. 2059 c.c. uno e bino: una interpretazione che non convince. Rassegna di Diritto Civile:775–783
Perlingieri P (2008) Conclusione dei lavori. In: Alcaro F, Fenga C, Moscati E, Pernice F, Tommasini R (eds) Valori della persona e modelli di tutela contro i rischi ambientali e genotossici. Esperienze a confronto. Firenze University Press, Firenze, pp 253–255
Perlingieri P (2009) L'onnipresente art. 2059 c.c. e la "tipicità" del danno alla persona. Rassegna di Diritto Civile, pp 520–529
Perlingieri P (2011) Le funzioni della responsabilità civile. Rassegna di Diritto Civile, pp 115–123
Perlingieri P (2015) Produzione, beni e benessere. Relazione conclusiva. In: V.Aa. (ed) Benessere e regole dei rapporti civili. Lo sviluppo oltre la crisi, in Atti del 9° Convegno Nazionale SISDiC, 8, 9 e 10 maggio 2014, Royal Continental Hotel - Napoli. Edizioni Scientifiche Italiane, Napoli, pp 509–519
Perlingieri P (2016) Persona, ambiente e sviluppo. In: Liber amicorum di Piero Alberto Capotosti, vol I. Cacucci, Bari, pp 535–550
Perlingieri P (2020a) Il diritto civile nella legalità costituzionale secondo il sistema italo-europeo delle fonti, 4th edn, vol IV, Attività e responsabilità. Edizioni Scientifiche Italiane, Napoli
Perlingieri P (2020b) Il diritto civile nella legalità costituzionale secondo il sistema italo-europeo delle fonti, I, Metodi e tecniche. Edizioni Scientifiche Italiane, Napoli
Pope Francesco (2015) Laudato sì, 24 May. https://vatican.va
Puleio G (2021) Rimedi civilistici e cambiamento climatico antropogenico. Persona e Mercato, pp 469–494
Quaranta, And (2024) Cop 28: quali risultati? Ambiente & Sviluppo 4:261–267
Salanitro U (2015) Giustizia e amministrazione al tempo della crisi: quale bilanciamento tra tutela della salute e salvaguardia dell'occupazione. In: V.Aa. (ed) Impresa e Mercato. Studi dedicati a Mario Libertini. Giuffré, Milano, pp 585–605
Salanitro U (2024) La responsabilità ambientale dopo la riforma costituzionale e la lotta al cambiamento climatico. Rivista di Diritto Civile 2:229–260
Savoia R (2017) No, una "bomba d'acqua" oramai non può più considerarsi un evento imprevedibile. Diritto & Giustizia 128:6
Tedoldi A (2024) ESG, next generation e rimedi collettivi (parte I), 29 January. https://quotidianogiuridico.it
Tozzi M (2023) La COP 28 non cura le cause della crisi. https://lastampa.it/editoriali/lettere-e-idee/2023/12/01/news/la_cop28_non_cura_le_cause_della_crisi-13901280/
Valaguzza S (2021) Liti strategiche: il contenzioso climatico salverà il pianeta? Diritto Processuale Amministrativo 2:293–334
Virgilio, Georgiche, III
Visintini G (2009) Cos'è la responsabilità civile. Fondamenti della disciplina dei fatti illeciti e dell'inadempimento contrattuale. Edizioni Scientifiche Italiane, Napoli
Worklimate (n.d.). https://worklimate.it
Zarro MC (2022) Danno da cambiamento climatico e funzione sociale della responsabilità. Edizioni Scientifiche Italiane, Napoli

Environment, Climate and Criticality of the Alleged Category of So-Called "Common" Goods

Giovanni Perlingieri

Abstract The chapter analyses the link between environmental and human protection, the criticality of the category of so-called "common" goods, the usefulness of the so-called "ecological interpretations of law". Through some case law, we focus on the aspects of resource management and legal standing for environmental damage. From this perspective it emerges the importance of the primacy of politics and principles.

1 Introduction

I intend to check whether the suggestion of certain doctrine to qualify the environment and climate as "common"[1] goods can lead to a qualitative and quantitative improvement of protection.

The article is inspired by the 17th National Conference of the Italian Society of Civil Law Scholars (S.I.S.Di.C.), on "Climate change, sustainability and civil relations", organized in the Aula Magna of the Faculty of Law of the University of Rome "La Sapienza", on 11, 12 and 13 January 2024. The essay is dedicated to the memory of Nicolò Lipari, President of the Session (n. III) "Climate change, people and market regulation".

[1] 1 See, Maddalena (2011), pp. 1 et seq.; Mattei (2011); Rodotà (2012a, reprint 2015), p. 113; Rodotà (2012b), p. 311; Mattei (2013), p. 111; Carapezza Figlia (2012), p. 535; Carapezza Figlia (2019), p. 185; Irti (2013), p. 381; Marinelli (2013), p. 406; Marinelli (2015); Cerulli Irelli and De Lucia (2014), p. 3; Mone (2014), p. 63; Francesca (2017), p. 473; Sirgiovanni (2017), p. 239; Fourth (2019), p. 933; Mazzarese (2020), p. 44; Lucarelli (2021), p. 21; Marcatajo (2022), p. 11. On the subject of so-called "common" goods *see* Perlingieri (2022), p. 137.

G. Perlingieri (✉)
Faculty of Law, Sapienza Università di Roma, Rome, Italy
e-mail: giovanni.perlingieri@uniroma1.it

G. Antonelli et al. (eds.), *Planetary Health - Laws, Policies and Science on the 'One Health' Approach*, https://doi.org/10.1007/978-3-031-90621-3_4

2 Insufficient "Definitions" of Environment and Climate. The Link Between Environmental, Climate and Human Protection

First, it is necessary to clarify the meaning of these concepts. With respect to the environment, doctrine has been divided between a pluralistic versus a unitary conception. Among the supporters of the first theory we count Massimo Severo Giannini who, before the introduction of law n. 349 of 1986 establishing the Ministry of the Environment, highlighted the use of the legislative concept of environment in at least three different meanings: one related to the landscape, one related to the protection of individual components (soil, water, air) and one used by the laws on urban planning.[2] The pluralistic conception has recently been reinforced by the doctrine that highlights the coexistence of different profiles, and sometimes conflicting, such as environmental hygiene, rational use of resources, as well as environmental protection as a source of knowledge and aesthetic-cultural emotion.[3] Though, another and most preferable approach emphasizes the link between environment and landscape, as a "way of being of the territory in its visible perception"[4] and, in particular, "the opportunity to consider the environment, from both a bioethical and legal point of view, in its unity, as a balanced relationship between man, fauna and flora, as an "overall ecosystem"".[5] Although they may seem different, these goods need to be protected in a uniform way because ecologism is an expression of personalism; protecting the environment, in all its aspects, it means that life is protected, as well as humans. This approach has been confirmed by several Supreme Courts (cfr. Norwegian Supreme Court in the Greenpeace case)[6] and by the Italian Constitutional Court, which has observed that biodiversity and the environment are at the same time the "object" of a "fundamental right-duty of the person" and a "fundamental interest of the community"[7] which, in a unified perspective, "includes the conservation, rational management and improvement of natural conditions (air, water, soil and land in all its components), the existence and preservation of terrestrial and marine genetic heritage, of all animal and plant species that live in it in its natural dimension and, ultimately, humans in all their manifestations".[8] According to the

[2] Giannini (1973), p. 15.

[3] Libertini (2015), p. 488.

[4] Predieri (1981), p. 506.

[5] Perlingieri (1991), p. 5. In a similar sense, Postiglione (1985), p. 32; and, most recently—after the reform of artt. 9 and 41 of the Italian Constitution, introduced by constitutional law 11 February 2022, n. 1—Tripodina (2023), p. 340.

[6] See https://climatecasechart.com/non-us-case/greenpeace-nordic-assn-v-ministry-of-petroleum-and-energy-ecthr/.

[7] Italian Constitutional Court 30 December 1987, no. 641, in Il Foro Italiano, (1988), with a commentary by Giampietro (1988).

[8] Italian Constitutional Court 28 May 1987, no. 210, Il Foro Italiano, I, (1988).

Italian Constitutional Court, "the environment constitutes a unitary intangible asset, although its several components, each of which may also constitutes separately the object of care and protection, but all together same parts of a unique system"[9] ("protection does not pursue abstract naturalistic or aesthetising purposes, but expresses the need for a natural habitat in which humans live and act, and which is necessary to the community to its citizens, according to widely felt values").[10] As a natural habitat instrumental to the development of the human person,[11] environmental protection cannot be reduced to the content and powers of subjective law only, but it can also find expression in powers of control, discretion and legitimate interests.

The recent amendment of articles 9 and 41 of the Italian Constitution—which in fact implicitly included already, at least under the systematic perspective of personalism and solidarism, the protection of the environment[12]—confirmed that "the civilization of a people is not measured by GDP, [...] but by the way in which it protects the dignity of the person, it ensures education and guarantees the protection of health and the environment".[13] Not by accident the World Health Organization defines the environment as "the set of physical, chemical, biological and social elements which exert an appreciable influence on the health and well-being of every human being".[14]

With respect to climate, in the lack of any legal definition, the doctrine has tried to provide a definition by referring, for its identification, to the most consolidated achievements of science. The concept of climate should be identified by referring to "two factual characteristics, full of legal implications".[15] On the one hand, it consists of a space-time variable of temperature, producing complex causal interactions of medium to long term, involving all forms of life.[16]

On the other hand, "it operates as an eco-systemic function of regulation of goods and services of nature, necessary also for human life and its socio-legal relations".[17] Thus climate is a "condition of the right, rather than its object",[18] which "regulates relationships instead of being regulated by them.[19]"

---

[9] Italian Constitutional Court 28 May 1987, no. 210, Il Foro Italiano, I, (1988).

[10] Perlingieri (1989), p. 136; Perlingieri (2016b), p. 323; Perlingieri (2020a), p. 334; Landi (2022), p. 95.

[11] Italian Constitutional Court 30 December 1987, no. 641, in Il Foro Italiano, I, (1988).

[12] Cfr. Perlingieri (2010), p. 403; recently Landi (2022), p. 52.

[13] Perlingieri (2020d), pp. 101 ff.; Perlingieri (2022), p. 156; Caterini (2018), p. 33.

[14] Cordini (2000), p. 437.

[15] Carducci (2021), p. 52.

[16] Carducci (2021), p. 52.

[17] Carducci (2021), p. 52.

[18] Carducci (2021), p. 53.

[19] Carducci (2021), p. 53.

This attempt of definition, however, ends up undermining the promotional function of law,[20] which postulates the need to regulate those human actions responsible of leading to an increase in global temperature—along with responsibilities of states, companies and citizens—in order to conform human activities to the principle of sustainability (not in its economic perspective, but in solidaristic dimension).[21]

Climate, as stated in Greenpeace case, is a factor that affects the environment, since its essential component, instrumental to the survival of the planet as well as of present and future generations.[22] Thus between climate and the protection of the value of humans there is an inseparable link.[23]

3 Diversity of Tools. Existence and Usefulness of the Category of Common Goods

According to the constitutional personalism and the solidarism, environment and climate are primary goods of life. Their protection, under an intergenerational[24] perspective, guarantees the collective enjoyment not only for the current generation, according to the traditional view of ownership or membership, but also, as Jhering[25] stated, for future generations.

In this perspective it is necessary to configure several tools of protection, taking into account the different interests involved in each individual case. Protection cannot be limited to the subjective right (which may also be expressed in powers of control, discretion and legitimate interests),[26] nor it can be reduced, under a remedies' perspective, to the prohibition of emissions (art. 844 of the Italian Civil Code) and to civil liability[27] (article 2043 of the Italian civil code), even if we try to re-read it in order to overrule the dogma of its compensatory nature (since compensation for environmental damage is mostly a punitive deterrent, since the damage to ecosystems, often irreversible, has already been produced in whole or in part).

Protection also requires the exploitation of other more effective remedies such as injunctive and enforcement tools, the control of legality and merit[28] of legal

---

[20] Bobbio (1969), p. 131. On the promotional function of public order in constitutional legality v. Perlingieri (2021c), p. 1382; Perlingieri and Zarra (2019).

[21] Perlingieri (2020b), p. 83; see also Perlingieri (2024), p. 167.

[22] Lipari (2024), p. 218.

[23] Perlingieri (2020c), p. 334; Perlingieri (2016b), p. 323; e already Perlingieri (1989), p. 136.

[24] Perlingieri (2020d), p. 116.

[25] Von Jhering (1906), p. 364.

[26] See Lipari (2023), p. 138.

[27] Piraino (2012), p. 466.

[28] Perlingieri (2003a), p. 16; Perlingieri (2003b), p. 396; Perlingieri (2003c), p. 423; Perlingieri (2020b), p. 99; Perlingieri (2014), p. 54; Martone (2017), p. 23; Minervini (2019), p. 13.

transactions, and consequently the re-reading of various provisions, such as article 2645 of the Italian civil code.

We may consider the bio-constraints on ownership with respect to agricultural land, in order to preserve its usefulness and avoid pollution, depletion and finally unusability of lands[29] due to the use of pesticides and fertilizers; or green procurement,[30] in the context of which ecological clauses are configured as instruments for safeguarding the value of the environment, and therefore of the person and his health, ultimately affecting the function of the contract. ts interpretation and execution.[31] Instead, considering the environment and climate as personality rights is quite reductive.[32] These are not autonomous rights, given the unitary value of all humans. The related protection must be examined from a perspective which, under the principle of constitutional solidarity, takes into account the duties[33] of both State and public bodies as well as of individuals and their association.[34] This is clearly the way to act in the interest of future generations.

We may wonder whether to achieve this goal it is necessary to refer, in addition to an autonomous category. The answer, which requires a re-reading of the classic theory of goods, seems to be negative. The legal discipline never depends on abstract qualifications only, in a typological perspective,[35] but on the interests and values related to goods, subjective situations and transactions.[36] Goods count for the ways and purposes of their use, not for their ownership.[37]

4 Limits of the Theory of the So-Called "Common" Goods: Resource Management and Environmental Liability

The doctrine of common goods, also proposed with reference to climate and atmosphere,[38] seems to be ambiguous, sometimes superfluous and even harbinger of abuses.[39]

---

[29] Vercellone (2023), p. 138; Perlingieri (2024), p. 167.

[30] See Irti (2017), p. 183; De Leonardis (2020), p. 62.

[31] In this sense, Patti (1979), p. 82; and, more recently, Leccese (2011); Leccese (2020), p. 36; Marcatajo (2021), p. 611, especially p. 615. Of right to the climate speak, recently, Franceschelli (2019), p. 457; Pisanò (2021), p. 261, especially p. 283; Pisanò (2022a), p. 21; Pisanò (2022b), p. 349.

[32] See Rimoli (2018), p. 39; Perlingieri (2020d), p. 116.

[33] Cfr. Piraino (2012), pp. 459, 466.

[34] Perlingieri, p. 512.

[35] Cfr. Piraino (2012), pp. 473, 491.

[36] Cfr. Míguez Núnez (2023), p. 431.

[37] Perlingieri (2022), p. 137; see Lipari (2013), pp. 134 and 128; Lipari (2023), p. 143.

[38] Nespor (2013), p. 672.

[39] Di Porto (2013), p. 72; Nivarra (2016), p. 43; Cicero and Perlingieri (2017), p. 803; Francesca (2017), p. 476; Perlingieri (2022), p. 139.

Ambiguous because it refers to very heterogeneous goods (such as air, water,[40] environment, territory,[41] coastline, cultural and artistic heritage, health, knowledge,[42] school, university, moon, work, justice, civic customs, democracy etc.),[43] and their "legal statutes deserve the most attention but are [...] to be reconstructed on a case-by-case basis, in relation to the nature and peculiarities of the function and of the interests specifically involved".[44]

Thus, the definition of common goods appears itself very uncertain, since it is considered "a category of being, of respect, of inclusion and quality [...] authentically relational, made up of relationships between individuals, communities, contexts and environment [...] ecological-qualitative and non-economic-quantitative category",[45] and "a common good, unlike private or public property [...] cannot be conceivened as a mere object, a tangible portion of the outside world. [...] The common good [...] exists only in a qualitative relationship".[46]

Beyond the doctrine of common goods, both in the national and European perspective (articles 41, 42 of the Italian Constitution; Articles. 345 Of the FEU Treaty and 17 Charter of Fundamental Rights EU, re-read according to articles 2, 3, 4 and 6 of the FEU Treaty, which protect the rights of the person) any kind of good, regardless of the ownership,[47] must have a social function.[48] In fact, "it is not the good itself that justifies a peculiar discipline, but it is the specific social function[49] to determining discipline, to establish powers, faculty and limits". While public property is a social function, "private property has a social function [...], since the interests of the owners is protected and recognized to perform both an individual and social function".[50]

The conception is also grounded on European law according to which private property does not neglect the protection of the "general interest" (art. 345 of the FEU Treaty and 17 Charter of Fundamental Rights) and, as confirmed by the EU Court of Justice, each right is connoted by "specific limits that act on the external sphere of law and limits inherent to its intimate content."[51] These limits change according to the nature of the interests involved.

---

[40] Cfr. Irti (2013), p. 381; Carapezza Figlia (2019), p. 193.

[41] Cfr. Maddalena (2014a, b), p. 19; Cerulli Irelli and De Lucia (2014), p. 10.

[42] Rodotà (2012a, reprint 2015), p. 130.

[43] See Marella (2012), p. 15.

[44] Perlingieri (2022), p. 142.

[45] Mattei (2011), p. 62.

[46] Mattei (2011), p. 52; Lucarelli (2013), p. 65.

[47] See Femia (2024), p. 197.

[48] Perlingieri (2022), p. 142.

[49] Perlingieri (2021b), p. 99.

[50] Perlingieri (2020c), p. 221; Perlingieri (2021b), p. 99; on common goods see. Carapezza Figlia (2011), p. 1072; Carapezza Figlia (2012), p. 535.

[51] *See* Court of Justice of the European Union, 27 January 2020, C-238/20.

The category of common goods appears also superfluous, not only because we already know the notions of *res in usu publico*, of *res communes omnium* and of public property, but mostly because any good has a social function: "good is not the *res*, good is the interest. The *res*, if there is one, constitutes a reason for the presence of interest."[52]

The ownership of an asset is also a relatively important profile;[53] even more relevant is the effectiveness of its use (exclusive, common, enlarged). A good of general interest will naturally have a different importance from another. Goods are relevant not for how ontologically and structurally appear, but for the interest and usefulness that they are destined to realize or satisfy.[54]

What is captured by law is not the intrinsic value of something, but its use.[55] If the good has protection in relation to its function and not to its qualification or its ownership, then the same romanistic idea of *res nullius* must be abandoned.[56] Even the distinction between goods of public property and goods of private property can become completely secondary.[57]

People are not interested on whether water, education, employment, culture, health care, transportation are private or public, but mostly in their quality and accessibility. There is no need to go beyond the opposition public/private to protect general interests. Rather, it is up to the jurist to determine from time to time whether, beyond the abstract qualification or the type of ownership, the goal has been reached, in accordance with the principle of effectiveness.[58]

The point is not the creation of a new category of goods, but rather the legal connection[59] between "things" and "people" which cannot be analyzed only in terms of ownership.

5 Further Critical Issues: Legal Standing for Compensation of Environmental Damage

The notion of common goods, replacing those of goods in common use or of public goods, it can also be counterproductive with respect to the topics of management, responsibility and legal standing. With the statement that common goods do not

[52] Perlingieri (2021c), p. 361.

[53] Lipari (2013), p. 134.

[54] Perlingieri (2021c), p. 357; Perlingieri (2022), p. 159.

[55] Míguez Núnez (2023), p. 452.

[56] See Piraino (2012), p. 476.

[57] See Perlingieri (2015), p. 515.

[58] Art. 24 of the Italian Constitution; UN Charter of 1948; art. 13 of the ECHR; art. 47 of the CFREU; art. 19 of the EU Treaty; art. 1 Italian administrative procedure code; art. 17 EU Regulation no. 679 of 2016.

[59] Mattei (2011), p. VII.

recognize "any other owner than those who have direct access to them",[60] the relation between these goods and public interest is neglected. In fact, regardless both existence and access,[61] this relation is essential to satisfy any kind of need (cfr. article 4 of the Italian Constitution). It is needed to avoid that "common property" or "common good" would mean "nobody's responsibility".

Natural resources are protected not only in the interest of local communities, but mostly in the interest of present and future generations.

The use of the notion of common good may carry some risks at the international level.

With respect to the extra-atmospheric space, the qualification of 'common' does not solve complex problems, nor clarifies the outcome of the balance between the protection of the common heritage of humanity (according to the definition of article 1 of the 1967 Treaty), the development of scientific and technical research and the rational exploitation of aerospace resources. The resolution of these problems cannot be separated from a complete regulation, yet to be implemented.[62]

The task of States should be the deployment of any power in order to protect the general interests related to these goods.

6 Conclusions

Like the notion of common good, the "ecological interpretations" of law seems ambiguous.

Certain "labels" such as ecological property and contract are sometimes even counter-productive, because they seem to admit anti-ecological contracts or property.

Both in the Italian and European legal system property rights (public and private) are physiologically limited.[63]

However, the primacy of politics on the economy is also entrusted to principles.

Hence the need to recognize, as positive law, a minimum body of higher principles, so that the primacy of politics is found through the application of shared general principles.

References

Bobbio N (1969) Sulla funzione promozionale del diritto. Rivista Trimestrale di Diritto e Procedura Civile, pp 1313–1329

---

[60] Libertini (2015), p. 489.

[61] Marella (2012), p. 20.

[62] Falcon (2016), p. 153; Zolea (2020), p. 13.

[63] Guarascio (2023), p. 73.

Carapezza Figlia G (2011) Premesse ricostruttive del concetto di beni comuni nella civilistica italiana degli anni Settanta. Rassegna di Diritto Civile, p 1072
Carapezza Figlia G (2012) Proprietà e funzione sociale. La problematica dei beni comuni nella giurisprudenza delle Sezioni unite. Rassegna di Diritto Civile, pp 535–549
Carapezza Figlia G (2019) Tecniche e ideologie nella disciplina delle acque. Dagli "usi di pubblico generale interesse" ai "beni comuni". In: Macario F, Addante A, Costantino D (eds) Studi in memoria di Michele Costantino, I. Edizioni Scientifiche Italiane, Napoli, pp 185–203
Carducci M (2021) Cambiamento climatico (dir. cost.). Digesto delle Discipline Pubblicistiche, VIII News Update. Wolters Kluwer, Milano, pp 51–74
Caterini E (2018) Sostenibilità e ordinamento civile. Per una riproposizione della questione sociale. Edizioni Scientifiche Italiane, Napoli
Cerulli Irelli V, De Lucia L (2014) Beni comuni e diritti collettivi. Politica del Diritto, pp 3–36
Cicero C, Perlingieri G (eds) (2017) Liber amicorum per Bruno Troisi, vol II. Edizioni Scientifiche Italiane, Napoli, pp 803–823
Cordini G (2000) I rapporti tra ordinamenti e l'interesse per l'ambiente. Il politico, pp 437–445
De Leonardis F (2020) L'uso strategico della contrattazione pubblica: tra GPP ["Green Public Procurement"] e obbligatorietà dei CAM. Rivista Quadrimestrale di Diritto dell'Ambiente, p 62
Di Porto A (2013) 'Res in usu publico' e 'beni comuni'. Il nodo della tutela. Giappichelli, Torino
Falcon M (2016) 'Res communes omnium'. Vicende storiche e interesse attuale di una categoria romana. In: Garofalo L (ed) I beni di interesse pubblico nell'esperienza giuridica romana, I. Jovene, Napoli, pp 107–164
Femia P (2024) Sentimento e moltitudine. Rudolf von Jhering tra interessi ideali e beni comuni. Il Mulino, Bologna
Francesca M (2017) Beni comuni e razionalità discreta del diritto. In: Perlingieri G, Fachechi A (eds) Ragionevolezza e proporzionalità nel diritto contemporaneo, I. Edizioni Scientifiche Italiane, Napoli, pp 473–492
Franceschelli F (2019) L'impatto dei cambiamenti climatici nel diritto internazionale. Editoriale Scientifica, Napoli
Giampietro F (1988) Il danno all'ambiente innanzi alla Corte costituzionale. Il Foro Italiano, I, col. 694
Giannini MS (1973) Ambiente: saggio sui diversi aspetti giuridici. Rivista Trimestrale di Diritto Pubblico, p 15
Guarascio C (2023) GreenEssence. La sostenibilità come motore del cambiamento economico. Il caso delle benefit corporation in Italia. In: Montesanti L (ed) Cambiamenti globali e sostenibilità. Edizioni Scientifiche Italiane, Napoli, pp 73–97
Irti N (2013) L'acqua tra beni comuni e concessioni (o la pluralità delle 'appartenenze'). Diritto e Società, 2013, pp 381–390
Irti C (2017) Gli "appalti verdi" tra pubblico e privato. Contratto e Impresa/Europa, pp 183–210
Landi R (2022) Risanamento ambientale e vicende della proprietà. Edizioni Scientifiche Italiane, Napoli
Leccese E (2011) Danno all'ambiente e danno alla persona. Franco Angeli, Milano
Leccese E (2020) L'ambiente: dal codice di settore alla costituzione, un percorso al contrario? (Riflessioni intorno ai progetti di legge per l'inserimento della tutela ambientale e dello sviluppo sostenibile tra i principi fondamentali della costituzione), pp 1–40. https://ambientediritto.it
Libertini M (2015) Persona, ambiente e sviluppo: ripensare la teoria dei beni. In: Aa.Vv. (ed) Benessere e regole dei rapporti civili. Lo sviluppo oltre la crisi, Atti del 9° Convegno Nazionale SISDiC in ricordo di Giovanni Gabrielli. Edizioni Scientifiche Italiane, Napoli, p 488
Lipari N (2013) Le categorie del diritto civile. Giuffré, Milano
Lipari N (2023) Vivere il diritto. A colloquio con Gabriele Carapezza Figlia, Vincenzo Cuffaro e Francesco Macario. Edizioni Scientifiche Italiane, Napoli
Lipari N (2024) Premesse per un diritto civile dell'ambiente. Rivista di Diritto Civile 2:209–228
Lucarelli A (2013) La democrazia dei beni comuni. Laterza, Bari

Lucarelli A (2021) Beni comuni, in Digesto delle Discipline Pubblicistiche, News Update, VIII, Torino, Utet, pp 21–28
Maddalena P (2011) L'ambiente e le sue componenti come beni comuni in proprietà collettiva della presente e delle future generazioni. Federalismi.it, pp 1–45
Maddalena P (2014a) Il territorio bene comune degli italiani. Donzelli Editore, Roma
Maddalena P (2014b) Il territorio bene comune degli italiani. Proprietà collettiva, proprietà privata e interesse pubblico. Donzelli editore, Roma
Marcatajo G (2021) La tutela dell'ambiente come diritto della persona. Rivista Giuridica dell'Ambiente, pp 611–639
Marcatajo G (2022) La riforma degli articoli 9 e 41 della Costituzione e la valorizzazione dell'ambiente. Rivista Giuridica dell'Ambiente, pp 1–15
Marella MR (2012) Introduzione. Per un diritto dei beni comuni. In: Marella MR (ed) Oltre il pubblico e il privato. Per un diritto dei beni comuni. Ombre Corte, Verona, pp 9–28
Marinelli F (2013) Usi civici e beni comuni. Rassegna di Diritto Civile, pp 406–422
Marinelli F (2015) Un'altra proprietà. Usi civici, assetti fondiari collettivi, beni comuni. Pacini Editore, Pisa
Martone I (2017) Il giudizio di meritevolezza. Questioni aperte e profili applicativi. Edizioni Scientifiche Italiane, Napoli
Mattei U (2011) Beni comuni. Un manifesto. Laterza, Roma-Bari
Mattei U (2013) I beni comuni fra economia, diritto e filosofia. Spazio filosofico, pp 111–116
Mazzarese T (2020) Diritti e beni comuni o fondamentali. Un binomio di riferimento per una risposta sistemica alle sfide dell'era globale. Ragion pratica, pp 441–460
Míguez Núnez R (2023) Beni a valore non prettamente economico: rilievi privatistici per un diritto "relazionale". Rassegna di Diritto Civile, pp 431–465
Minervini E (2019) La "meritevolezza" del contratto. Una lettura dell'art. 1322, comma 2, c.c. Giappichelli, Torino
Mone D (2014) La categoria dei beni comuni nell'ordinamento giuridico italiano: un paradigma per la lettura del regime dei beni pubblici alla luce della Costituzione. Rassegna di Diritto Pubblico Europeo online
Nespor S (2013) Tragedie e commedie nel nuovo mondo dei beni comuni. Rivista Giuridica dell'Ambiente, pp 665–685
Nivarra L (2016) Quattro usi di "beni comuni" per una buona discussione. Rivista Critica del Diritto Privato, p 43
Patti S (1979) La tutela civile dell'ambiente. Cedam, Padova
Perlingieri P (1989) Spunti in tema di tutela dell'ambiente. Legalità e Giustizia, pp 136–139
Perlingieri P (1991) Presentazione. In: Perlingieri P (ed) Il danno ambientale con riferimento alla responsabilità civile. Edizioni Scientifiche Italiane, Napoli, pp 5–7
Perlingieri P (2003a) I mobili confini dell'autonomia privata. In: Perlingieri P (ed) Il diritto dei contratti fra persona e mercato. Problemi del diritto civile. Edizioni Scientifiche Italiane, Napoli, pp 13–22
Perlingieri P (2003b) In tema di tipicità e atipicità nei contratti. In: Perlingieri P (ed) Il diritto dei contratti fra persona e mercato. Problemi del diritto civile. Edizioni Scientifiche Italiane, Napoli, pp 389–413
Perlingieri P (2003c) Nuovi profili del contratto. In: Perlingieri P (ed) Il diritto dei contratti fra persona e mercato. Problemi del diritto civile. Edizioni Scientifiche Italiane, Napoli, pp 415–439
Perlingieri G (2010) Sul giurista che come "il vento non sa leggere". Rassegna di Diritto Civile, pp 385–413
Perlingieri G (2014) Il controllo di "meritevolezza" degli atti di destinazione ex art. 2645 ter. Il Foro Napoletano, pp 54–100
Perlingieri P (2016b) Persona, ambiente e sviluppo. In Pennasilico M (ed) Contratto e ambiente. L'analisi "ecologica" del diritto contrattuale, Atti del Convegno di Bari, 22–23 ottobre 2015. Edizioni Scientifiche Italiane, Napoli, pp 321–341

Perlingieri P (2020a) Il diritto civile nella legalità costituzionale secondo il sistema italo-europeo delle fonti, vol IV, Attività e responsabilità, 4ª ed. Edizioni Scientifiche Italiane, Napoli
Perlingieri P (2020b) Il diritto civile nella legalità costituzionale secondo il sistema italo-europeo delle fonti, vol III, Situazioni soggettive, 4ª ed. Edizioni Scientifiche Italiane, Napoli
Perlingieri P (2020c) Il diritto civile nella legalità costituzionale secondo il sistema italo-europeo delle fonti, vol I, Metodi e tecniche, 4ª ed. Edizioni Scientifiche Italiane, Napoli
Perlingieri G (2020d) "Sostenibilità", ordinamento giuridico e "retorica" dei diritti. A margine di un recente libro. Il Foro Napoletano, pp 101–118
Perlingieri P (2021b) Relazione conclusiva. In: Capobianco E, Perlingieri G, D'Ambrosio M (eds) Circolazione e teoria dei beni. Edizioni Scientifiche Italiane, Napoli, p 359
Perlingieri G (2021c) In tema di ordine pubblico. Rassegna di Diritto Civile, pp 1382–1436
Perlingieri G (2022) Criticità della presunta categoria dei beni c.dd. "comuni". Per una "funzione" e una "utilità sociale" prese sul serio. Rassegna di Diritto Civile, pp 137–164
Perlingieri G (2024) In tema di "sostenibilità". Teoria e Prassi del Diritto, pp 167–178
Perlingieri G, Zarra G (2019) Ordine pubblico interno e internazionale tra caso concreto e sistema ordinamentale. Edizioni Scientifiche Italiane, Napoli
Piraino F (2012) Sulla nozione di bene giuridico in diritto privato. Rivista Critica del Diritto Privato, pp 459–494
Pisanò A (2021) Il diritto al clima. Una prima concettualizzazione. L'Ircocervo, pp 261–283
Pisanò A (2022a) Il diritto al clima. Il ruolo dei diritti nei contenziosi climatici europei. Edizioni Scientifiche Italiane, Napoli
Pisanò A (2022b) La responsabilità degli Stati nel contrasto al cambiamento climatico tra obbligazione climatica e diritto al clima. Etica & Politica, pp 349–366
Postiglione A (1985) Ambiente: suo significato giuridico unitario. Rivista Trimestrale di Diritto Pubblico, pp 32–50
Predieri A (1981) Paesaggio. In: Enciclopedia del Diritto, XXI. Giuffré, Milano, p 506
Pugliatti S (1962) Beni e cose in senso giuridico. Giuffré, Milano
Rimoli F (2018) Sulla retorica dei diritti. Mucchi Editore, Modena
Rodotà S (2012a, reprint 2015) Il diritto di avere diritti. Laterza, Roma-Bari
Rodotà S (2012b) Beni comuni: una strategia globale contro lo human divide. In: Marella MR (ed) Oltre il pubblico e il privato. Per un diritto dei beni comuni. Ombre Corte, Verona, pp 311–332
Sirgiovanni B (2017) Dal diritto sui beni comuni al diritto ai beni comuni. Rassegna di Diritto Civile, pp 229–246
Tripodina C (2023) La tutela dell'ambiente nella Costituzione italiana: tra interessi delle generazioni future e responsabilità della generazione presente. Rivista Quadrimestrale di Diritto dell'Ambiente, pp 332–349
Vercellone A (2023) Proprietà ecologica e art. 2645 ter c.c.: il "bio-vincolo". Rassegna di Diritto Civile, pp 131–158
von Jhering R (1906) Geist des römischen Rechts auf den verschiedenen Stufen seiner Entwicklung, vol 3. Druck und Verlag von Breitkopf und Härtel, Leipzig
Zolea S (2020) Esplorazione spaziale e nuove forme di appartenenza. Cardozo Electr Law Bull:1–41

Reverse Correlation Between Allergic Pollens and Fungi Amounts Annually with Allergic Sensitization Rate to Them

Young-Jin Choi, Kyung-Suk Lee, and Jae-Won Oh

Abstract A growing number of people have recently contracted allergic diseases caused by pollen because of climate change. The seasonal variations in pollens have occurred with the increased sensitization rate to pollens in children as well as in adults. Allergic plants which are rapidly proliferating have emerged as a dangerous element for allergic children.

Plants causing allergies are strongly proposed to be due not only to the rapid proliferation of weeds and trees but also to air pollution such as greenhouse gases as well as increased temperatures resulting from climate change. As an altered climate affects the range of allergenic plant species and the length of the pollen season, a number of studies have shown that increased CO_2 levels and atmospheric temperature raise pollen concentration as well as increased allergenicity of pollens with duration of the extended pollen season. Weather variables with CO_2 levels are among the main factors affecting phenology and pollen production. Weather patterns influence the movement and dispersion of all aeroallergens depending on atmospheric stability.

Otherwise, the atmospheric fungal concentrations decreased annually for 25 years. The intensity of precipitation during the summer and autumn increased, but the number of days with precipitation decreased throughout the seasons. Even though there was an increase in precipitation, it was concentrated over a short period of time. These changes may have augmented dry season duration, which reduced the sporulation period for fungi. Allergenic fungal sporulation could decrease with climate changes in future. There is shown reverse phenomenon between increased

Y.-J. Choi · K.-S. Lee · J.-W. Oh (✉)
Division of Allergy, Respiratory Diseases, Department of Pediatrics, College of Medicine, Hanyang University, Seoul, South Korea

Department of Pediatrics, Hanyang University Guri Hospital, Guri, South Korea
e-mail: jaewonoh@hanyang.ac.kr

G. Antonelli et al. (eds.), *Planetary Health - Laws, Policies and Science on the 'One Health' Approach*, https://doi.org/10.1007/978-3-031-90621-3_5

allergic pollens and decreased fungi amounts annually with allergic sensitization rate to them.

It is yet unknown whether complex interactions with pollens, fungi, meteorological variables, and air pollutants in the changing environment. It is crucial to eliminate the associated risk for human health in future and take appropriate measures to reduce it.

1 Introduction

The climate is constantly changing over time and is dependent on incoming solar radiation, outgoing thermal radiation and the composition of the atmosphere on earth. Changes are also occurring in the amount, intensity, and frequency of precipitation and the occurrence of extreme events such as heat waves, droughts, and hurricanes (Smith et al. 2009, p. 4133). Many prevalent human diseases are linked to climate fluctuations, from cardiovascular mortality and respiratory illnesses to altered transmission of infectious diseases and malnutrition from crop failures (Patz et al. 2005, p. 310). Climate change is a constant process that affects allergies. The influence of a changing climate on allergic disease has generated interest among the public and the scientific community. Impacts on pollen and fungal allergy may be one of the important consequences of climate change on human health. The present knowledge is taken from experimental and epidemiological studies on the relationship between allergic respiratory diseases and environmental factors, such as meteorological variables, airborne allergens, and air pollution, while studies on respiratory allergy triggered by climate change are still to evaluate (Peñuelas and Filella 2001, p. 793; Running 2008, p. 652).

2 Pollen and Fungi as Aeroallergen

For a particle to be clinically significant as an aeroallergen, it must be buoyant, present in significant numbers, and allergenic such as pollen. In general, the insect-pollinated plants do not produce appreciable amounts of airborne pollen, as opposed to wind-pollinated plants, which, by necessity, produce particles that travel for miles. Pollen size and weight as aeroallergens as they fall in the 20–60 μm range of most particulate matter and their allergenic constituents are proteins with a molecular weight between 10,000 and 40,000 Da (Fig. 1) (Stewart 2000, p. 1107; Oh 2023, p. 155).

Protective mechanisms in the nasal mucosa and tracheobronchial passages remove most of the larger particles, so only those 5 μm or smaller reach the alveoli of the lungs. The development of asthma after pollen exposure is enigmatic because pollen grains are deposited in the upper airways as a result of their large particle size. Asthma due to pollen may be caused by the inhalation of pollen debris that is

Fig. 1 Major allergenic pollen grains and plants in Korea. (**a**) Alder, (**b**) Birch, (**c**) Oak, (**d**) Ragweed, (**e**) Mugwort, (**f**) Japanese hop (Permitted from Korean Pollen Allergy Institution and Foundation)

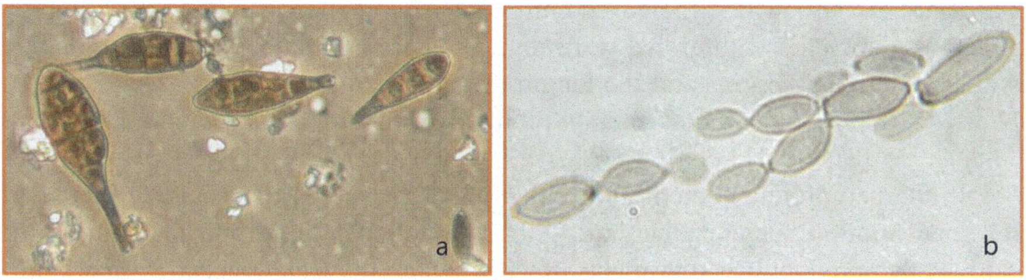

Fig. 2 Major allergenic fungal spores in Korea. (**a**) *Alternaria alternata*, (**b**) *Cladosporium herbarum* (x400) (Permitted from Korean Pollen Allergy Institution and Foundation)

small enough to access the bronchial tree. These are considerations in the pathogenesis of allergic rhinitis and asthma (Ziska 2020, p. 771; D'Amato et al. 2007, p. 976).

Fungi may be unicellular; syncytial, with many nuclei not divided into separate cells; or multicellular, with nuclei separated by septa. Fungi reproduce by producing spores, many of which are adapted for airborne dispersal. Spores may be produced by either meiosis or mitosis. Among allergenic fungi, most spores produced by mitosis are formed on differentiated hyphae or conidia. These are associated with the asexual (anamorphic) stage of the life cycle (Taylor et al. 1999, p. 197; Hibbett et al. 2007, p. 509). Fungal spores are highly allergenic, and may be more numerous than pollen grains in the air (Fig. 2). They secrete enzymes into their surroundings and absorb the breakdown products of enzyme action. Some of these enzymes are well-known allergens (Crameri et al. 2009, p. 99). Once inhalation, atmospheric fungal propagules may trigger mast cell bound IgE-mediated allergic response in atopic individuals who may subsequently develop symptoms of allergic respiratory diseases, such as allergic rhinitis, allergic conjunctivitis and allergic asthma (Soeria-Atmadja et al. 2010, p. 125).

3 Pollen, Fungi and CO_2

The greatness of climate change related to rising allergic diseases will be affected by how greenhouse gas mitigation strategies are followed up on. CO_2 is the predominant greenhouse gas (Curtis 1996, p. 127). Human activities have increased the natural concentration of CO_2 in our atmosphere, increasing earth's natural greenhouse effect (Bazzaz 1990, p. 167; NOAA/ESRL 2017). In 1870, prior to the industrial revolution, CO_2 was 280 parts per million (ppm) to 409.92 on January 1, 2019, according to a new analysis of air samples collected by National Oceanic and Atmospheric Administration (NOAA)'s Global Monitoring Division (GMD) (Albertine et al. 2014, e111712; Drake et al. 1997, p. 609). CO_2 uptake is the main biogeochemical effect. The increase of CO_2 as well as the air temperature in temperate regions can induce increased carbon uptake and prolonged growing periods. In general, plants are expected to grow more rapidly during the period of global warming. An extended plant activity season increases CO_2 uptake and thus decreases the current rise of atmospheric CO_2 concentration and its influence on the greenhouse effect (Albertine et al. 2014, e111712). The extended plant activity also further increases the total annual emission of biogenic volatile organic compounds (Drake et al. 1997, p. 609). These increased emissions may also contribute to the complex processes associated with global warming. An altered climate will affect the range of allergenic plant species and the length of the pollen season, and elevated atmospheric CO_2 levels may increase plant productivity and pollen production affecting plant ecology (Jablonski et al. 2002, p. 9; LaDeau and Clark 2001, p. 292).

The allergen protein content in a pollen grain can increase because of increased CO_2 concentration, such that the same number of pollen grains differs in their allergenicity. Pollen production experiments of trees by altering CO_2 concentration are limited by canopy size and exposure time. Kim et al. used open-top CO_2 chambers to maintain CO_2 concentrations at ambient (400 ppm), ×1.4 (560 ppm), and ×1.8 (720 ppm) levels. The production of pollen grains and the allergen Que. a 1 by oak trees in the chambers were quantitatively assessed (Fig. 3). Total pollen counts per tree of the ×1.4 and ×1.8 levels significantly increased to 353% and 1299%, respectively, compared to the ambient level. The allergenic protein content at the ×1.4 and ×1.8 levels showed a significant increase to 12% and 11%, respectively (Kim et al. 2018, p. 1587). Otherwise, in the glass chamber study for ragweed pollen, there were significantly different concentrations of major allergen Amb a 1 among the chambers as followed results; 1.88 ng Amb a 1/μg pollen in the 400 ppm, 4.44 ng/μg pollen in the 600 ppm (2.36 times vs 400 ppm, $p < 0.032$), and 5.36 ng/μg pollen in the 1000 ppm (2.85 times vs 400 ppm, $p < 0.025$) (Fig. 4) (Choi et al. 2018, p. 278).

Mycorrhizal fungi form vast underground networks found in soils on every continent of Earth, from forests to grasslands to croplands. These fungi, both living and after they die, form a part of soil organic matter and play an essential role in the soil's ability to retain carbon — roughly 75% of Earth's terrestrial carbon is stored in soil (Soudzilovskaia et al. 2019, p. 5077). Experiments have also shown that increasing CO_2 increases fungal spore production. In this study, with regards to

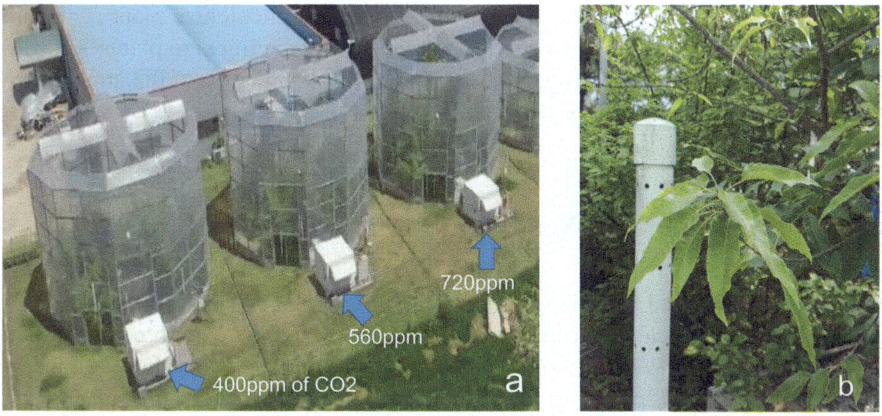

Fig. 3 (**a**) Open Top Chambers with each labeled with CO_2 concentration: 400 ppm, 560 ppm, and 720 ppm at the National Institute of Forest Science in Suwon, Korea and (**b**) its internal view showing white pole with holes with a CO_2 injection nozzle and plants (Permitted from Kim KR, Oh JW, et al. Int J Biometeorol 62:1587–94, 2018)

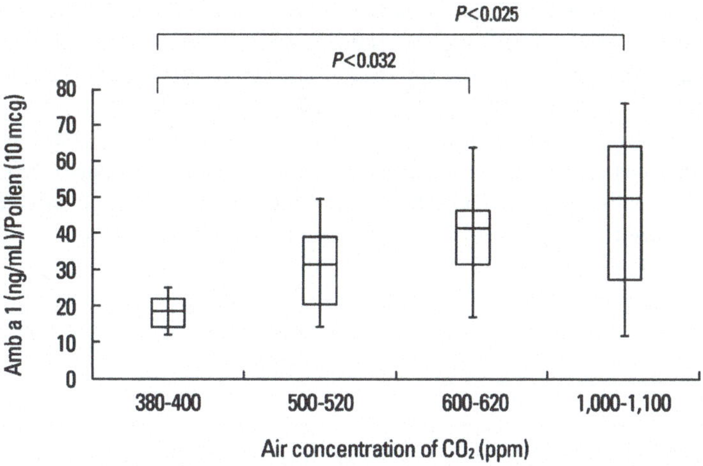

Fig. 4 Comparison of ragweed major allergen (Amb a 1) concentration in each CO2 chamber ($n = 20$). 1.88 ng Amb a 1/μg pollen in the 400 ppm, 4.44 ng/μg pollen in the 600–620 ppm (2.36 times vs 400 ppm, $p < 0.032$), and 5.36 ng/μg pollen in the 1000–1100 ppm (2.85 times vs 400 ppm, $p < 0.025$) (Permitted from Choi YJ, et al. Allergy Asthma Immunol Res. 2018;10(3):278–82)

climate variables, both CO_2 and maximum temperature reported statistically significant increasing trends (Klengel et al. 2005, p. 2021; Mogensen et al. 2006, p. 103). For trees and several molds, the average number of active weeks shows a strong increasing trend over the years (Paudel et al. 2021, p. 12816).

Some rust fungi (Uredinales or Pucciniales) are negatively affected by higher temperatures and increased concentrations of atmospheric CO_2. These fungi and plants have a mutually beneficial relationship. The fungi colonize plant roots and extend far beyond their hosts to provide nutrients that are critical to plants' growth. Plants return the favor with sugars made from CO_2 absorbed from the atmosphere during photosynthesis (Helfer 2014, p. 770).

4 Pollen, Fungi and Weather

Impacts on pollen and fungi may be one of the important consequences of weather change on human health. Weather change can influence the occurrence of pollen and fungus allergy. Weather conditions, including rainfall, atmospheric temperature, humidity, wind speed and direction, may alter the concentrations of pollens, fungi, and other aeroallergens which can subsequently influence the occurrence of allergic diseases.

A number of studies of the effect of climate change on the distribution of allergenic pollen and fungi have focused typically on the analysis of their observed concentrations and their regression relationships with meteorological factors. The onset, duration and intensity of the pollen season vary annually. Weather variables and greenhouse gases are among the main factors affecting phenology and pollen production (Bonofiglio et al. 2013, p. 877; Ziello et al. 2012, e34076; Zhang et al. 2013, p. 64; Sleeter et al. 2013, p. 733; Russell and McBride 2003, p. 64). Weather play a key role in the production, release and bioavailability of pollen- and fungi-derived allergens (O'Driscoll et al. 2005, p. 4; Beaubien and Freeland 2000, p. 53). In addition, weather patterns influence the production, movement and dispersion of pollen and fungi in the atmosphere through air temperature, winds, rainfall, and precipitation depending on atmospheric stability (García-Mozo et al. 2014, p. 103; Parmesan and Yohe 2003, p. 37). Plant species require a certain amount of heat to complete their development; then air temperature plays a key role, together with other factors, such as day length, water and nutrient availability, and soil type. Increasing worldwide evidence demonstrates that the timing of life cycle events of a large number of species has responded to the increase in air temperature. Plants discriminate day from night by means of photoreceptors, that is, pigments that capture different wavelengths that may promote or inhibit flowering. These pigments synchronize biological activities with the day and night cycle (Menzel 2000, p. 76).

The climate and weather directly determine the start and end of pollination, which defines the duration of the pollen season. Plants may respond to day lengths exceeding a critical threshold in late spring or early summer, securing time for seed maturation. Increasing light intensity during springtime and early summer is likely to influence flowering in summer-flowering plants and modify the effect of the photoperiod (Latorre 1999, p. 131). The recent study based on massive data also illustrates a clear positive correlation between recent global warming and an increase in the seasonal duration and amount of pollen for multiple allergenic plant species on a decadal basis for the northern hemisphere. The atmospheric pollen concentrations collated and analyzed here reflect spatial and temporal changes in allergenic plant

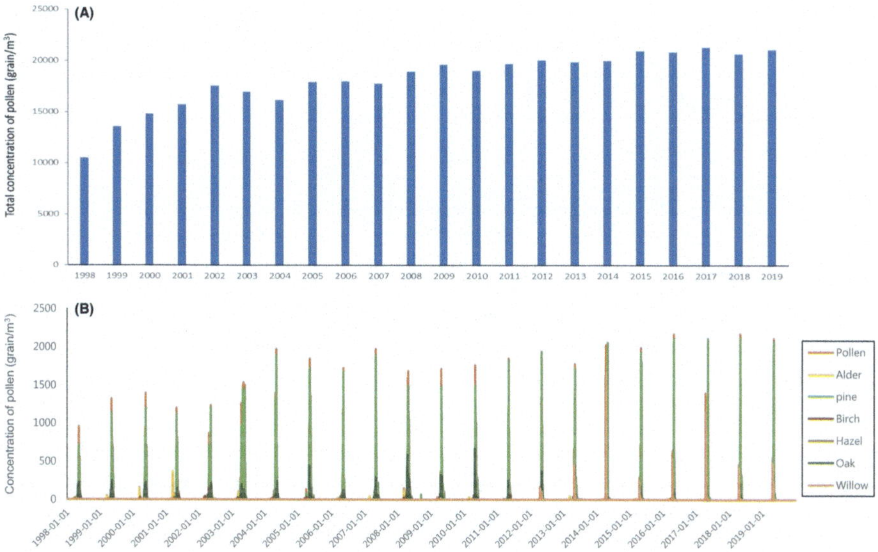

Fig. 5 The trend in the total concentrations (**a**) and the peak concentrations (**b**) of the studied pollens in the Seoul metropolitan area during the 22 years of the study. Both the total and the peak concentrations of pollens increased gradually over time (Permitted from Lee KS, et al. Ped Allergy Immunol 2021;32(5):872–79)

species for that location, these data indicate that recent climatic changes, through temperature, are in fact already affecting pollen amounts as well as season duration and timing in the northern hemisphere (Ziska et al. 2019, e124). Thus, Climate change could increase both the concentration of allergenic pollens and the duration of the pollen season (Figs. 5 and 6) (Lee et al. 2021, p. 872). In turn, allergic symptoms may be dependent on the exposure pollen level and duration. The weather may influence the occurrence of allergic diseases such as asthma and allergic rhinitis (D'Amato et al. 2015, p. 25). These observed changes have immediate and future health implications, particularly for allergic diseases such as allergic rhinitis and asthma. Future climate projections also suggest extension in the duration of the pollen season and an estimated rising in sensitization rate to allergenic plants (Lee et al. 2021, p. 872; Solomon and LaRocque 2019, p. 209: D'Amato et al. 2016, p. 434; Hegerl et al. 2007, p. 663; Watts et al. 2018, p. 2479).

Atmospheric spore concentrations are influenced by a wide array of environmental, meteorological factors and various interspecies interactions. The number of allergic fungal spores underlies seasonal variations, with climatic factors influencing the spectrum of fungal species and their amounts in the atmospheric environment (Barnes 2019, p. 439; D'Amato et al. 2020, p. 2219). Atmospheric fungi are generally known to cause allergic reactions in late Summer and early Fall (Atkinson et al. 2006, p. 580; Twaroch et al. 2015, p. 205). Coupled with increasing exposures to air pollutants and other aeroallergens, weather effects may also play a role on the distribution and concentration of allergenic fungi. The species and amounts of fungi

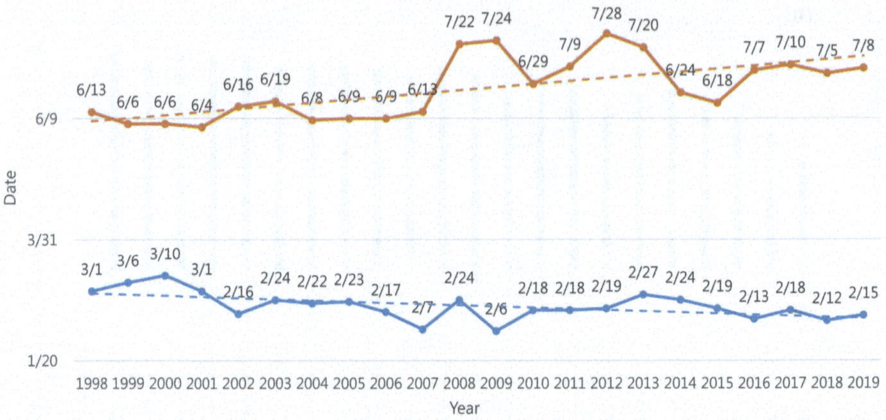

Fig. 6 The start and end dates of the pollen season for the major allergenic tree pollens in the Seoul metropolitan area during the past 22 years. In 1998, the start and end dates were March 5 and June 7, respectively, and the duration of the pollen season was 94 days. In 2019, the start and end dates were February 15 and July 5, respectively, and the duration of the pollen season was 140 days (Permitted from Lee KS, et al. Ped Allergy Immunol 2021;32(5):872–79)

present in the air depend on regional and seasonal variations in vegetation, air temperature, relative humidity, and seasonal changes (D'Amato and Spieksma 1995; Montealegre et al. 2004, p. 51). Humidity is a crucial factor for sporulation of fungi. However, knowledge on the determinants of spore production, seasonal dynamics, sources, and dispersion is still need to evaluate. There is no convincing alternative information supported by observational evidences.

Over the past 25 years, Korea has recorded a long-term trend of higher precipitation with annual precipitation increasing by +17.71 mm per decade. However, the number of precipitation days has decreased correspondingly by 21.2 days (Fig. 7). The amount and intensity of precipitation during the summer and autumn increased significantly. On the other hand, the number of days with precipitation decreased throughout the season. Even though there was an increase in precipitation, it was concentrated over a short period of time. These changes in precipitation trends may have extended the duration of dry season, which shortened the sporulation period for fungi requiring moisture. Eventually the concentration of fungal spores may have reduced as the exposure to spores decreased (Fig. 8) (Choi et al. 2023, p. 825).

5 Allergic Sensitization Rate to Pollen and Fungi

Hypersensitivity to pollens is one of the most common aspects of the allergic disease, leading to allergic rhinitis, allergic conjunctivitis, and asthma (Oh 2022, p. 168). An interaction between the exposure to pollen and being sensitized to pollen allergens was detected that the effect was less pronounced in individuals who

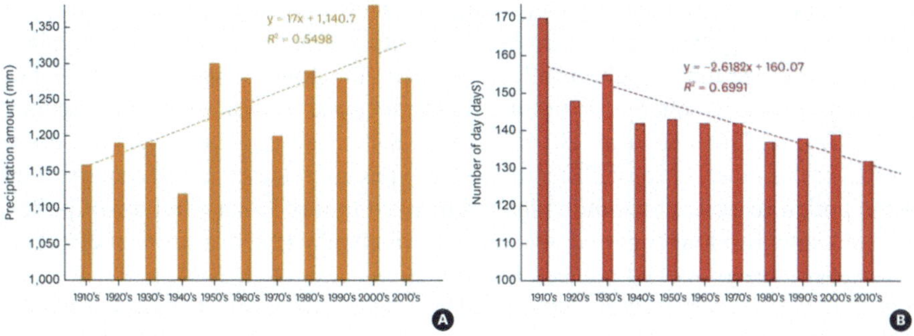

Fig. 7 Longitudinal precipitation trends per decade in Korea over the last 100 years. (**a**) Precipitation; (**b**) Number of precipitation days. The annual precipitation increased by 135.4 mm, while the number of precipitation days decreased by 21.2 days. The annual precipitation over the last 100 years increased by +17.71 mm per decade (Permitted from Choi YJ, et al. Allergy Asthma Immunol Res. 2023;15(6):825–36)

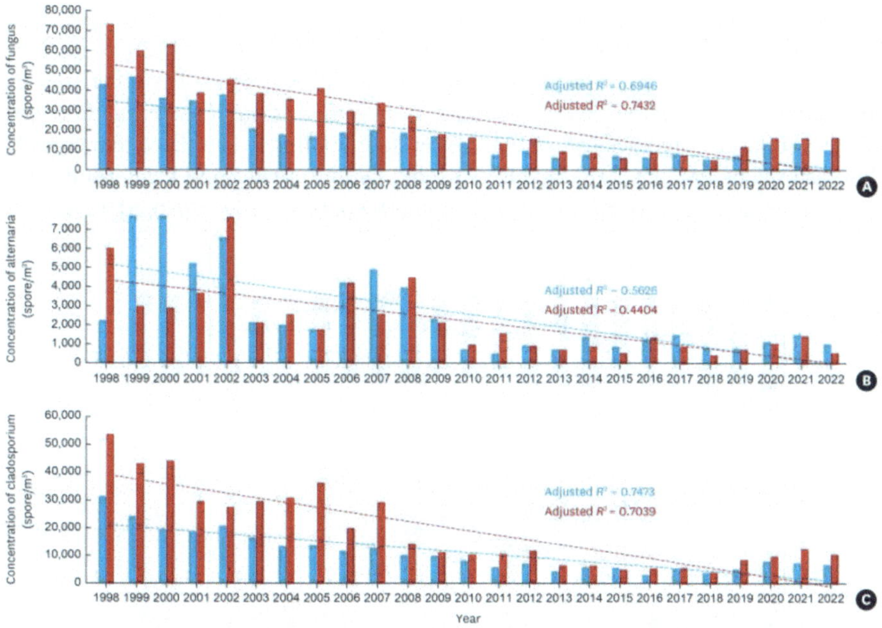

Fig. 8 Changes in the annual concentrations of fungal spores of different types (A: total fungal spores, B: Alternaria, C: Cladosporium) in Seoul (blue) and Guri (red), Korea over a period of 25 years (1998–2022). The total annual atmospheric fungal concentrations decreased in both areas (Permitted from Choi, et al. Allergy Asthma Immunol Res. 2023;15(6):825–36)

were sensitized to one type of pollen, compared with who are sensitized to all types of pollen. There is the different threshold level of allergic sensitization among plant species. Sensitization to multiple allergens, including pollens, has been reported to lead to more serious allergic symptoms and to increase the risk of asthma (Schoos et al. 2019, p. 716; Lee et al. 2019, p. 1542; Kim et al. 2012, p. 350; Park et al. 2016). Furthermore, because different pollen types are released at different periods of the pollen season, atopic individuals sensitized to more than one type of pollen are experiencing a longer period of exposure, which may lead to a stronger effect on airway hyperreactivity and reduced lung function (Choi et al. 2021, p. 97; Bielory et al. 2012, p. 485; Shea et al. 2008, p. 443). Among a number of studies, 10–30% of the global population is affected by allergic rhinitis due to seasonal pollen exposure, and multiple interrelated, potential consequences of climate change (Reid and Gamble 2009, p. 458). A longer pollen season and larger amounts of pollen may enhance human exposure to allergenic pollens, potentially leading to an increase in allergic sensitization rate with rising pollen sensitization rate in children have recently been reported (Fig. 9) (Lee et al. 2021, p. 872).

In a few studies of fungal allergy there were presented the exploring the hospitalizations, or asthma symptoms revealed associations with *Alternaria,* and *Cladosporium* (Katotomichelakis et al. 2016, e51; Twaroch et al. 2015, p. 205; Randriamanantany et al. 2010, p. 368; Olsen et al. 2020, p. 141). The other species taxa reported associations with *Coprinus, Aspergillus, Penicillium*, and *Botrytis*. This is similar to outcomes from the previous study finding to be related to increased child asthma emergency department visits and *Cladosporium* and Deuteromycetes (Caillaud et al. 2018a, p. 837; Caillaud et al. 2018b, p. 170137; Bush 2016, p. 433;

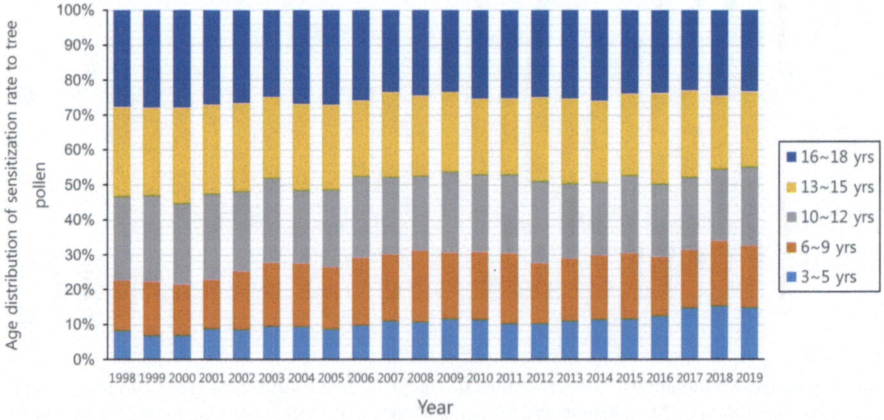

Fig. 9 Age distribution of sensitization rates to the major pollens in the Seoul metropolitan area during the study period. Between 1998 and 2019, pollen-sensitization rates increased in children <10 years of age (1998: 3–5 years: 8.3%, 6–9 years: 14.4%, 10–12 years: 24.0%, 13–15 years: 25.4%, and 16–18 years: 27.6%; 2019:3–5 years: 10.7%, 6–9 years: 17.7%, 10–12 years: 22.5%, 13–15 years: 23.9%, and 16–18 years: 25.3%) (Permitted from Lee KS, et al. Ped Allergy Immunol 2021;32(5):872–79)

Baxi et al. 2016, p. 396). *Cladosporium* spp. and *Alternaria* spp. spores are omnipresent in the environment and are two dominating genera in the total air spores. The threshold level of spore necessary to elicit allergic symptoms in atopic individual is not known and varies between different species (Rodríguez-Rajo et al. 2005, p. 497).

Alternaria alternata is mainly an outdoor fungus that generally grows on vegetation, but, this species can also be found indoors. Indoor *Alternaria alternata* concentrations are typically impacted by the current outdoor concentrations and often produces large brown spores that are a well-known cause of allergy and asthma (Pulimood et al. 2007, p. 120). Sensitization rates to fungi vary according to different studies, but allergy to *Alternaria alternata* seems to affect roughly 5% of people and is related to respiratory allergic diseases (Lyons et al. 2011, p. 301; Skjøth et al. 2012, p. 11107). *Cladosporium* is one of the major sources of inhaled fungal allergy, which means allergic individuals sensitized to *C. herbarum* may experience symptoms after breathing in its spores (García-Mozo et al. 2014, p. 103). *Cladosporium* is among the top three most common indoor airborne fungi along with *Penicillium* and *Aspergillus*. In a recent study, the sensitization rates to *Alternaria* and *Cladosporium* decreased annually in both areas over the study period. The allergic sensitization rates to *Alternaria* were 3.5% and 0.2% in 1998 and 2022, respectively. The allergic sensitization rates to *Cladosporium* were 4.4% and 0.2% in 1998 and 2022, respectively in Seoul, Korea (Fig. 10) (Choi et al. 2023, p. 825).

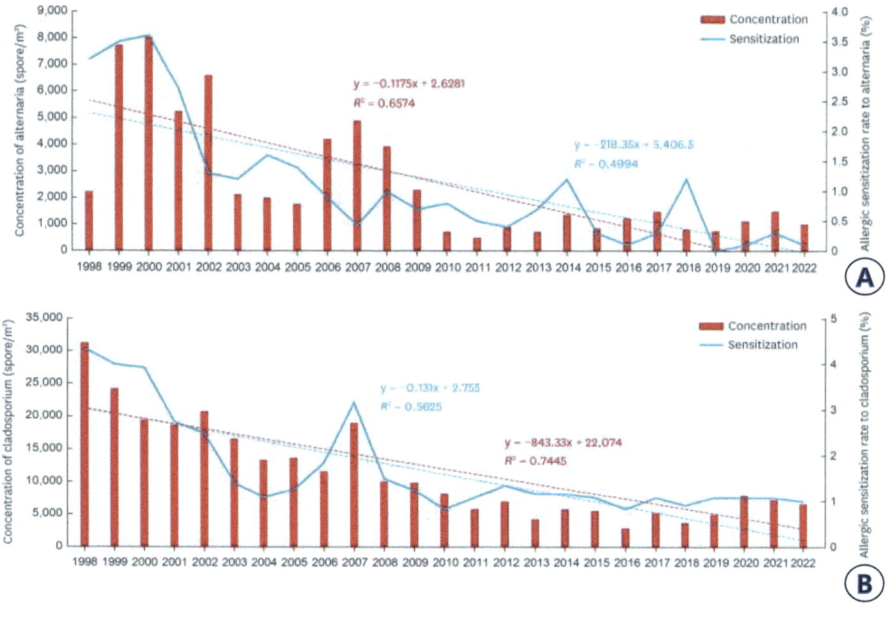

Fig. 10 Correlations between atmospheric concentrations of *Alternaria* (**A**) and *Cladosporium* spores (**B**) (red bar) and allergic sensitization rates (blue line) to *Alternaria* (**A**) and *Cladosporium* (**B**) in Seoul, Korea (Permitted from Choi, et al. Allergy Asthma Immunol Res. 2023;15(6):825–36)

6 Reverse Trend Between Allergic Pollens and Fungi Concentration with Sensitization Rate

In recent study, allergenic tree pollen concentration increased and the sensitization rate to pollens increased annually (Lee et al. 2021, p. 872). Otherwise, the concentrations and the sensitization rates to *Alternaria* and *Cladosporium* decreased annually in the Seoul metropolitan for 25 years (Choi et al. 2023, p. 825). Olsen's 26 years study presented that despite increasing annual temperature and decreasing relative humidity, they found a decreasing trend for seasonal and annual peak concentrations of *Alternaria* and *Cladosporium* (Olsen et al. 2020, p. 141) This pattern was dissimilar, even reverse to that of fungi allergens. In this period, the amount and intensity of precipitation during the summer and autumn increased significantly, and the number of days with precipitation decreased throughout the seasons. Even though there was an increase in precipitation, it was concentrated over a short period of time. These changes in precipitation trends may have augmented dry season duration, which reduced the sporulation period for fungi requiring moisture. It is suspicious whether this reverse trend is due to the impact of climate change or not. Probably the concentration and distribution of pollens and fungi are significantly affected by environmental changes due to climate change such as drought, atmospheric dryness, and prolonged dry seasons. Therefore, prolonged monitoring periods and further large-scale studies are required to confirm causality and to evaluate the impact of climate change in future (Fig. 11).

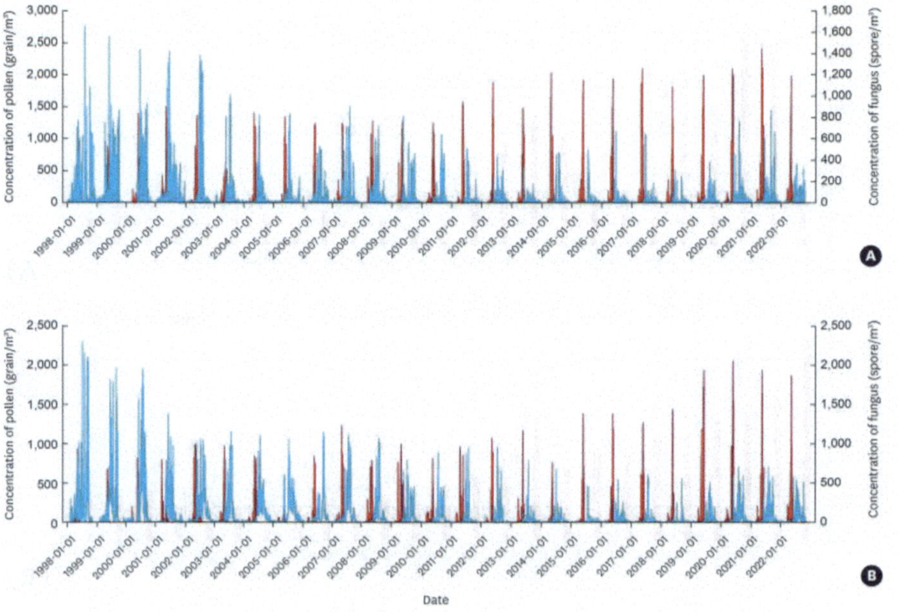

Fig. 11 Comparison of the annual concentrations of pollen grains (red) and fungal spores (blue) in Seoul (**a**) and Guri (**b**) for the past 25 years (1998–2022). In contrast to atmospheric fungal concentration, pollen concentration increased annually during the same period (Permitted from Choi, et al. Allergy Asthma Immunol Res. 2023;15(6):825–36)

References

Albertine JM, Manning WJ, DaCosta M, Stinson KA, Muilenberg ML, Rogers CA (2014) Projected carbon dioxide to increase grass pollen and allergen exposure despite higher ozone levels. PLoS One 9:e111712

Atkinson RW, Strachan DP, Anderson HR, Hajat S, Emberlin J (2006) Temporal associations between daily counts of fungal spores and asthma exacerbations. Occup Environ Med 63:580–590

Barnes C (2019) Fungi and atopy. Clin Rev Allergy Immunol 57:439–438

Baxi SN, Portnoy JM, Larenas-Linnemann D, Phipatanakul W, Environmental Allergens Workgroup (2016) Exposure and health effects of fungi on humans. J Allergy Clin Immunol Pract 4:396–404

Bazzaz FA (1990) The response of natural ecosystems to the rising global CO_2 levels. Annu Rev Ecol Syst 21:167–196

Beaubien EG, Freeland HJ (2000) Spring phenology trends in Alberta, Canada: links to ocean temperature. Int J Biometeorol 44:53–59

Bielory L, Lyons K, Goldberg R (2012) Climate change and allergic disease. Curr Allergy Asthma Rep 12(6):485–494

Bonofiglio T, Orlandi F, Ruga L, Romano B, Fornaciari M (2013) Climate change impact on the olive pollen season in Mediterranean areas of Italy: air quality in late spring from an allergenic point of view. Environ Monit Assess 185:877–890

Bush RK (2016) Fungal sensitivity: new insights and clinical approaches. J Allergy Clin Immunol Pract 4:433–434

Caillaud D, Cheriaux M, Martin S, Segala C, Dupuy N, Evrard B et al (2018a) Short term effect of outdoor mould spore exposure on prescribed allergy medication sales in Central France. Clin Exp Allergy 48:837–845

Caillaud D, Leynaert B, Keirsbulck M, Nadif R, Mould ANSES Working Group (2018b) Indoor mould exposure, asthma and rhinitis: findings from systematic reviews and recent longitudinal studies. Eur Respir Rev 27:170137

Choi YJ, Oh HR, Oh JW, Kim KR, Kim BJ, Baek WG (2018) Chamber and field studies demonstrate differential amb a 1 contents in common ragweed depending on CO_2 levels. Allergy Asthma Immunol Res 10:278–282

Choi YJ, Kee KS, Oh JW (2021) The impact of climate change on pollen season and allergic sensitization to pollens. Immunol Allergy Clin N Am 41:97–109

Choi YJ, Lee KS, Jeong JH, Kim KH, Yang S et al (2023) Annual change in fungal concentrations and allergic sensitization rates to *Alternaria* and *Cladosporium* in Korea during the period 1998–2022. Allergy Asthma Immunol Res 15(6):825–836

Crameri R, Zeller S, Glaser AG, Vilhelmsson M, Rhyner C (2009) Cross-reactivity among fungal allergens: a clinically relevant phenomenon? Mycoses 52:99–106

Curtis PS (1996) A meta-analysis of leaf gas exchange and nitrogen in trees grown under elevated carbon dioxide. Plant Cell Environ 19:127–137

D'Amato G, Spieksma FT (1995) Aerobiologic and clinical aspects of mould allergy in Europe. Allergy 50:870–877

D'Amato G, Cecchi L, Bonini S, Nunes C, Annesi-Maesano I, Behrendt H et al (2007) Allergenic pollen and pollen allergy in Europe. Allergy 62:976–990

D'Amato G et al (2020) The effects of climate change on respiratory allergy and asthma induced by pollen and mold allergens. Allergy 75:2219–2228

D'Amato G, Holgate ST, Pawankar R et al (2015) Meteorological conditions, climate change, new emerging factors, and asthma and related allergic disorders. A statement of the world allergy organization. World Allergy Organ J 8(1):25

D'Amato G, Vitale C, Lanza M, Molino A, D'Amato M (2016) Climate change, air pollution, and allergic respiratory diseases: an update. Curr Opin Allergy Clin Immunol 16(5):434–440

Drake BG, Gonzalez-Meler MA, Long SP (1997) More efficient plants: a consequence of rising atmospheric CO_2? Annu Rev Plant Physiol Plant Mol Biol 48:609–639

García-Mozo H, Yaezel L, Oteros J, Galán C (2014) Statistical approach to the analysis of olive long-term pollen season trends in southern Spain. Sci Total Environ 473–474:103–109

Hegerl GC, Zwiers FW, Braconnot P et al (2007) Understanding and attributing climate change. In: Solomon S, Qin D, Manning M et al (eds) Climate change 2007: the physical science basis. Contribution of the Working Group I to the Fourth Assessment Report of the Intergovernamental Panel on Climate Change. Cambridge University Press, Cambridge, pp 663–746

Helfer S (2014) Rust fungi and global change. New Phytol 201:770–780. https://doi.org/10.1111/nph.12570

Hibbett DS, Binder M, Bischoff JF, Blackwell M, Cannon PF, Eriksson OE et al (2007) A higher-level phylogenetic classification of the fungi. Mycol Res 111(Pt 5):509–547

Jablonski LM, Wang X, Curtis PS (2002) Plant reproduction under elevated CO_2 conditions: a meta-analysis of reports on 79 crop and wild species. New Phytol 156:9–26

Katotomichelakis M, Nikolaidis C, Makris M, Proimos E, Aggelides X, Constantinidis TC et al (2016) *Alternaria* and *Cladosporium* calendar of Western Thrace: relationship with allergic rhinitis symptoms. Laryngoscope 126:E51–E56

Kim JH, Oh JW, Lee HB, Kim SW, Kang IJ, Kook MH et al (2012) Changes in sensitization rate to weed allergens in children with increased weeds pollen counts in Seoul metropolitan area. J Korean Med Sci 27:350–355

Kim KR, Oh JW, Woo SY, Seo YA, Choi YJ, Kim HS et al (2018) Does the increase in ambient CO2 concentration elevate allergy risks posed by oak pollen? Int J Biometeorol 62:1587–1594

Klengel T, Liang WJ, Chaloupka J, Ruoff C, Schroppel K, Naglik JR et al (2005) Fungal adenylyl cyclase integrates CO_2 sensing with cAMP signaling and virulence. Curr Biol 15:2021–2026

LaDeau SL, Clark JS (2001) Rising CO_2 levels and the fecundity of forest trees. Science 292:95–98

Latorre F (1999) Differences between airborne pollen and flowering phenology of urban trees with reference to production, dispersal and interannual climate variability. Aerobiologia 15:131–141

Lee SW, Yon DK, James CC, Lee S, Koh HY, Sheen YH et al (2019) Short-term effects of multiple outdoor environmental factors on risk of asthma exacerbations: age-stratified time-series analysis. J Allergy Clin Immunol 144:1542–1550.e1

Lee KS, Kim KH, Choi YJ, Yang S, Kim CY et al (2021) Increased sensitization rates to tree pollens in allergic children and adolescents and a change in the pollen season in the metropolitan area of Seoul, Korea. Pediatr Allergy Immunol 32(5):872–879

Lyons TW, Wakefield DB, Cloutier MM (2011) Mold and *Alternaria* skin test reactivity and asthma in children in Connecticut. Ann Allergy Asthma Immunol 106:301–307

Menzel A (2000) Trends in phenological phases in Europe between 1951 and 1996. Int J Biometeorol 44:76–81

Mogensen EG, Janbon G, Chaloupka J, Steegborn C, Fu MS, Moyrand F, Klengel T et al (2006) *Cryptococcus neoformans* senses CO_2 through the carbonic anhydrase Can2 and the adenylyl cyclase Cac1. Eukaryot Cell 5:103–111

Montealegre F, Meyer B, Chardon D, Vargas W, Zavala D, Hart B et al (2004) Comparative prevalence of sensitization to common animal, plant and mould allergens in subjects with asthma, or atopic dermatitis and/or allergic rhinitis living in a tropical environment. Clin Exp Allergy 34:51–58

NOAA/ESRL (2017) Trends in atmospheric carbon dioxide. http://www.esrl.noaa.gov/gmd/ccgg/trends/2017

O'Driscoll BR, Hopkinson LC, Denning DW (2005) Mold sensitization is common amongst patients with severe asthma requiring multiple hospital admissions. BMC Pulm Med 5:4

Oh JW (2022) Pollen allergy in a changing planetary environment. Allergy Asthma Immunol Res 14(2):168–181

Oh JW (2023) Pollen allergy. In: Oh JW (ed) Pollen allergy in a changing world. A guide to scientific understanding and clinical practice, 2nd edn. Springer Nature, New York, pp 155–161

Olsen Y, Skjøth CA, Hertel O, Rasmussen K, Sigsgaard T, Gosewinkel U (2020) Airborne *Cladosporium* and *Alternaria* spore concentrations through 26 years in Copenhagen, Denmark. Aerobiologia 36:141–157. https://doi.org/10.1007/s10453-019-09618-7

Park HJ, Lee JH, Park KH, Kim KR, Han MJ, Choe H et al (2016) A six-year study on the changes in airborne pollen counts and skin positivity rates in Korea: 2008–2013. Yonsei Med J 57:714–720

Parmesan C, Yohe G (2003) A globally coherent fingerprint of climate change impacts across natural systems. Nature 421:37–42

Patz JA, Campbell-Lendrum D, Holloway T, Foley JA (2005) Impact of regional climate change on human health. Nature 438(7066):310–317. https://doi.org/10.1038/nature04188

Paudel B, Chu T, Chen M, Sampath V, Prunicki M, Nadeau KC (2021) Increased duration of pollen and mold exposure are linked to climate change. Sci Rep 11:12816. https://doi.org/10.1038/s41598-021-92178-z

Peñuelas J, Filella I (2001) Phenology. Responses to a warming world. Science 294(5543):793–795. https://doi.org/10.1126/science.1066860

Pulimood TB, Corden JM, Bryden C, Sharples L, Nasser SM (2007) Epidemic asthma and the role of the fungal mold *Alternaria alternata*. J Allergy Clin Immunol 120:610–617

Randriamanantany ZA, Annesi-Maesano I, Moreau D, Raherison C, Charpin D, Kopferschmitt C et al (2010) *Alternaria* sensitization and allergic rhinitis with or without asthma in the French Six Cities study. Allergy 65:368–375

Reid CE, Gamble JL (2009) Aeroallergens, allergic disease, and climate change: impacts and adaptation. EcoHealth 6(3):458–470

Rodríguez-Rajo FJ, Iglesias I, Jato V (2005) Variation assessment of airborne Alternaria and Cladosporium spores at different bioclimatical conditions. Mycol Res 109:497–507

Running SW (2008) Climate change. Ecosystem disturbance, carbon, and climate. Science 321(5889):652–653. https://doi.org/10.1126/science.1159607

Russell WH, McBride JR (2003) Landscape scale vegetation-type conversion and fire hazard in the San Francisco Bay area open spaces. Landsc Urban Plan 64:201–208

Schoos AM, Jelding-Dannemand E, Stokholm J, Bønnelykke K, Bisgaard H, Chawes BL (2019) Single and multiple time-point allergic sensitization during childhood and risk of asthma by age 13. Pediatr Allergy Immunol 30:716–723

Shea KM, Truckner RT, Weber RW et al (2008) Climate change and allergic disease. J Allergy Clin Immunol 122(3):443–453

Skjøth C, Sommer J, Frederiksen L, Gosewinkel Karlson U (2012) Crop harvest in Denmark and Central Europe contributes to the local load of airborne *Alternaria* spore concentrations in Copenhagen. Atmos Chem Phys 12:11107–11123

Sleeter BM et al (2013) Land-cover change in the conterminous United States from 1973 to 2000. Glob Environ Chang 23:733–748

Smith JB, Schneider SH, Oppenheimer M, Yohe GW, Hare W, Mastrandrea MD, Patwardhan A, Burton I, Corfee-Morlot J, Magadza CH, Füssel HM, Pittock AB, Rahman A, Suarez A, van Ypersele JP (2009) Assessing dangerous climate change through an update of the Intergovernmental Panel on Climate Change (IPCC) "reasons for concern". Proc Natl Acad Sci USA 106(11):4133–4137. https://doi.org/10.1073/pnas.0812355106

Soeria-Atmadja D, Onell A, Borga A (2010) IgE sensitization to fungi mirrors fungal phylogenetic systematics. J Allergy Clin Immunol 125:1379–1386.e1

Solomon CG, LaRocque RC (2019) Climate change - a health emergency. N Engl J Med 380(3):209–211

Soudzilovskaia NA, van Bodegom PM, Terrer C, van't Zelfde M, McCallum I, McCormack ML et al (2019) Global mycorrhizal plant distribution linked to terrestrial carbon stocks. Nat Commun 10:5077. https://doi.org/10.1038/s41467-019-13019-2

Stewart GA (2000) The molecular biology of allergens. In: Busse WW, Holgate ST (eds) Asthma and rhinitis. Blackwell Scientific, Boston, pp 1107–1142

Taylor J, Jacobson D, Fisher M (1999) The evolution of asexual fungi: reproduction, speciation and classification. Annu Rev Phytopathol 37:197–246

Twaroch TE, Curin M, Valenta R, Swoboda I (2015) Mold allergens in respiratory allergy: from structure to therapy. Allergy Asthma Immunol Res 7:205–220

Watts N, Amann M, Arnell N et al (2018) The 2018 report of the lancet countdown on health and climate change: shaping the health of nations for centuries to come. Lancet 392(10163):2479–2514
Zhang Y, Isukapalli S, Bielory L, Georgopoulos P (2013) Bayesian analysis of climate change effects on observed and projected airborne levels of birch pollen. Atmos Environ 68:64–73
Ziello C et al (2012) Changes to airborne pollen counts across Europe. PLoS One 7:e34076
Ziska LH (2020) An overview of rising CO2 and climatic change on aeroallergens and allergic diseases. Allergy Asthma Immunol Res 12:771–782
Ziska LH, Makra L, Harry SK, Bruffaerts N, Hendrickx M, Coates F et al (2019) Temperature-related changes in airborne allergenic pollen abundance and seasonality across the northern hemisphere: a retrospective data analysis. Lancet Planet Health 3:e124–ee31

California's Wildfire Crisis and the Future of Planetary Resilience

Jessica Yu and Eeshan Chaturvedi

Abstract This chapter examines the escalating threats of climate change-driven wildfires to human and planetary health. Exacerbated by rising temperatures and droughts, wildfires degrade air quality, damage ecosystems, and endanger public health while also displacing communities, destroying infrastructure, and incurring substantial economic losses. Wildfires further exacerbate climate change by releasing greenhouse gases, creating a destructive feedback loop. Using California as a case study, we assess wildfire resilience through a planetary health framework, integrating strategies across the biosphere, society, and economy. Key interventions span the biosphere, including prescribed burns and forest management; society, by providing equitable resources for community mitigation and fire-resistant infrastructure; and the economy, incorporating sustainable industrial practices and utility infrastructure investments. Aligned with the United Nations Sustainable Development Goals (SDGs), our approach emphasizes proactive, interdisciplinary strategies to protect communities and ecosystems. By combining traditional and modern wildfire management practices, we provide actionable solutions to mitigate wildfire risks and foster long-term planetary resilience in an era of climate change.

The authors would also like to acknowledge the kind feedback on Sections 4 and 5 provided by Cassandra Jurenci and Eric Macomber, Climate and Energy Policy Program.

J. Yu (✉)
Stanford University, Stanford, CA, USA
e-mail: jesyu@stanford.edu

E. Chaturvedi
Emmett Interdisciplinary Program in Environment and Resources, Stanford University, Stanford, CA, USA
e-mail: eeshan@stanford.edu

G. Antonelli et al. (eds.), *Planetary Health - Laws, Policies and Science on the 'One Health' Approach*, https://doi.org/10.1007/978-3-031-90621-3_6

1 Introduction

Over the period from 1979 to 2022, the duration of fire seasons expanded across a quarter of the world's vegetated areas, leading to a nearly one-fifth increase in the average global fire season length.[1] Wildfires are fast becoming a global menace that has caused immense damage to the environment. In recent years, significant attention has been drawn to the pressing issues of large-scale and intense wildfires, including their origins, causes, and repercussions.[2] The increased frequency and intensity of these fires are a result of several factors: climate change, which has intensified hot and dry conditions; past fire suppression policies, leading to almost a century of fuel buildup (woody debris and other vegetation); and insufficient fuel management, resulting in the removal of too little of this accumulated material.[3] Regardless of the ignition source, these wildfires have led to a transformation in fire regimes extending from local to global scales.[4] In the US, this has been most evident in California and other western states, but wildfire activity has increased across the nation and around the world.[5] The number of large wildfires and areas burned by wildfires each year in the western United States has increased significantly over the last 30 years.[6] This trend is expected to continue, even in areas historically characterized by long fire return intervals.[7] While many studies project continued increases in wildfire activity, the extent and nature of these changes can vary by region and depend on multiple factors, including climate change, land management practices, and vegetation dynamics.[8]

The increasing threat of wildfires underscores the importance of planetary health, a concept popularized by a 2015 Rockefeller Foundation–Lancet Commission. Planetary health considers not only the effects of environmental factors on population health but also the consequences of threats to the global environmental systems that support humankind - how are we harming the earth, and how does that harm humanity?[9] Wildfires cause wide-ranging adverse health impacts, either directly due to the fire and smoke-related air pollution, or indirectly because of subsequent water and soil pollution, as well as from the stress and trauma associated with displacement, loss of property, and loss of life. Acute and chronic illnesses resulting from wildfires can occur in humans, often triggered by inflammation or DNA damage, leading to respiratory and cardiovascular diseases, cancer, mental health issues, and

[1] Intergovernmental Panel on Climate Change (2022).

[2] Pandey et al. (2023), p. 100431.

[3] Miller et al. (2020), pp. 101–109.

[4] Bilbao et al. (2020), pp. 435–496.

[5] Bowman et al. (2020a), pp. 500–515. https://doi.org/10.1038/s43017-020-0085-3.

[6] Westerling (2016), p. 20150178.

[7] Liu and Wimberly (2016), pp. 65–75.

[8] Andela et al. (2023), p. 137. https://doi.org/10.1038/s43248-023-00363-4.

[9] Chang et al. (2021), pp. 2029–2031. https://doi.org/10.1161/CIRCULATIONAHA.121.054946.

preterm birth in infants.[10] The impacts of climate change, particularly wildfires, are a cause for serious concern, with planetary health equity being called in crisis.[11]

We aim to, through this chapter, integrate the interconnected challenges of wildfires and planetary health, demonstrating how wildfire policies in California affect both human lives and the natural systems upon which humans and wildlife depend. To effectively preserve planetary health in the context of wildfires, three core tenets—biosphere, society, and economy—must be considered. Prioritizing the hierarchical and intertwined nature of these tenets will increase the success of policies designed to mitigate and manage wildfires and their planetary impacts. Focusing on strategic approaches in California, this chapter analyzes the current state of wildfire policy and contributes to the existing literature on managing this planetary health crisis. Each of these interventions is discussed in greater detail below.

2 Case Study: A Decade of Devastating Wildfires in California

Over the past decade, California has faced a series of unprecedented and catastrophic wildfires, marking some of the most significant and destructive events in the state's history.[12] The 2018 wildfire season was particularly devastating, with over 1.8 million acres burned.[13] The Camp Fire alone, which ravaged the town of Paradise, resulted in 85 fatalities and destroyed more than 18,800 structures, making it the costliest natural disaster of that year due to insured losses.[14] In the same year, more than 459,000 acres were lost to the Mendocino Complex Fire, setting a record as the largest wildfire in California's history at the time.[15] The 2020 wildfire season shattered previous records with over 4.3 million acres burned, driven by prolonged droughts, extreme heat, and strong winds. During the same season, the August Complex Fire eclipsed the Mendocino Complex Fire to become the largest recorded wildfire in California, burning over 1 million acres.[16] Most recently, in January 2025, the Greater Los Angeles region experienced devastating wildfires that burned over 57,000 acres, destroyed more than 18,000 structures, and resulted in at least 30 fatalities.

---

[10] Akdis and Nadeau (2022), pp. 651–652. https://doi.org/10.1038/s41577-022-00779-1.

[11] Friel (2023), p. 100195.

[12] Turco et al. (2023), p. e2213815120; Ayars et al. (2023), p. e2312909120.

[13] CAL FIRE. "2018 Fire Season." https://www.fire.ca.gov/incidents/2018/ (last visited Sept. 1, 2024).

[14] Nauslar et al. (2018).

[15] CAL FIRE. "Top 20 Largest California Wildfires." https://www.fire.ca.gov/media/4jandihh/top20_acres.pdf (accessed Sept. 1, 2024).

[16] CAL FIRE. "2020 Fire Season." https://www.fire.ca.gov/incidents/2020/ (last visited Sept. 1, 2024).

The impacts of these wildfires on communities and the environment have been profound. Economic costs have soared, with damages in 2018 alone approaching $28 billion.[17] Pacific Gas and Electric (PG&E), the state's utility company, filed for bankruptcy after being held liable for failing to maintain its equipment, which was found to have ignited several fires.[18] Health impacts have also been severe, with smoke exposure contributing to respiratory problems and increased mortality rates.[19] Ecologically, the fires have devastated habitats, affecting over 500 vertebrate species, and resulted in the loss of significant portions of the world's mature giant sequoias during the 2020 Castle Fire.[20] These challenges have been further exacerbated by climate change, which has extended fire seasons and increased the frequency and intensity of wildfires.[21] In what is being called the most devastating economic loss in Californian history, the more recent 2025 wildfires in Los Angeles county caused estimated damages ranging from $95 billion to $164 billion, including property and capital losses.[22] Without significant intervention, the increasing threat of wildfires poses a continued risk to vulnerable communities and ecosystems in California.

In this case study, we delve into the devastating impacts of wildfires on the population, society, and ecosystems of California. First, we compare California's wildfire impacts with those in other regions of North America, highlighting how region-specific environmental and community factors influence the severity and outcomes of these events. Next, we propose how incorporating core tenets of a planetary health framework could improve wildfire management strategies in California and other wildfire-prone regions around the world, ultimately aiming to mitigate the escalating risks posed by wildfires.

3 Contextualizing California Wildfires in North America

California's wildfires stand out as some of the most destructive and costly in the United States.[23,24] This devastation is largely due to the state's unique combination of environmental factors, including prolonged droughts, extreme heat, and

[17] Wang et al. (2021), p. 252. https://doi.org/10.1038/s41893-020-00645-2.

[18] Kousky et al. (2018). https://ludwig.guru/s/if+available.

[19] Reid et al. (2016), p. 1334. https://doi.org/10.1289/ehp.1409277.

[20] Stephens et al. (2020), p. 354. https://doi.org/10.1002/fee.2181.

[21] Williams et al. (2019), p. 892. https://doi.org/10.1029/2018EF001049.

[22] Li and Yu (2025).

[23] Cova et al. (2023), p. 120620. https://doi.org/10.1016/j.foreco.2022.120620.

[24] Wang et al. (2021), pp. 252–260. https://doi.org/10.1038/s41893-020-00645-2.

extensive development in fire-prone regions[25,26,27], particularly the wildland-urban interface, where human development meets with vegetative fuels like forests, grassland, or shrublands.[28] While other U.S. regions also contend with wildfires, experiencing increasing fire size and impacts on communities, the scale and overall economic impact are more severe in California given its population density.[29] In Oregon, the 2021 wildfire season resulted in devastating consequences, including over 1.1 million acres burned and over 400 homes lost.[30] That same year, California experienced over 2.5 million acres burned and over 3600 structures damaged or destroyed.[31] The economic impact was estimated between $70 billion and $90 billion for the entire U.S., with $45 billion to $55 billions of those damages occurring in California due to its larger population and economy.[32] While California's experience underscores the devastating potential of wildfires, the growing threat across numerous states necessitates comprehensive mitigation and management strategies to address the escalating risks exacerbated by climate change.

Although California experiences more frequent fires, the 2023 wildfire season in Canada demonstrated the growing global nature of wildfire challenges, with fires that were unprecedented in size and environmental impact.[33] The wildfires in Canada significantly affected air quality across North America, including cities as far as New York, which experienced some of the worst air quality in decades.[34,35] These fires also contributed to 24% of global fire carbon emissions that year,[36] which was equivalent to more emissions than 7 of the 10 largest emitting nations in 2022.[37] While both California and Canada face substantial economic costs due to wildfires, the nature and scope of the impact in each sector may differ. In California, economic damages are significantly tied to property losses and firefighting

25 Turco et al. (2023), p. e2213815120. https://doi.org/10.1073/pnas.2213815120.

26 Ayars et al. (2023), p. e2312909120. https://doi.org/10.1073/pnas.2312909120.

27 State of California (2021). https://resources.ca.gov/-/media/CNRA-Website/Files/Initiatives/California_Wildfire_and_Forest_Resilience_Action_Plan.pdf.

28 Radeloff et al. (2018), p. 3314. https://doi.org/10.1073/pnas.1718850115.

29 Li, Zhiyun, and William Yu. "USA: Economic impact of the Los Angeles wildfires." PreventionWeb, February 12, 2025. https://www.preventionweb.net/news/economic-impact-los-angeles-wildfires (accessed February 19, 2025).

30 United States Department of Agriculture, Forest Service. *2021 Wildfire Summary*. May 2022. https://www.fs.usda.gov/Internet/FSE_DOCUMENTS/fseprd1103226.pdf.

31 California Department of Forestry and Fire Protection (CAL FIRE), "Incidents Archive 2021," https://www.fire.ca.gov/incidents/2021 (accessed February 19, 2025).

32 Kasischke et al. (2010), p. 1313. https://doi.org/10.1139/X10-078.

33 Safford et al. (2022), p. 962.

34 Cedeño Laurent et al. (2024), pp. 14753–14763. https://doi.org/10.1021/acs.est.4b03444.

35 Chen et al. (2025), p. 120941. https://doi.org/10.1016/j.atmosenv.2024.120941.

36 COPERNICUS ATMOSPHERE MONITORING SERV. "Canada produced 23% of the global wildfire carbon emissions for 2023." https://atmosphere.copernicus.eu/copernicus-canada-produced-23-global-wildfire-carbon-emissions-2023 (accessed Sept. 1, 2024).

37 Byrne et al. (2024).

expenses,[38] driven by high population density and development in fire-prone areas, although the effects are not limited to these areas. For instance, the economic fallout from the Los Angeles wildfires alone could reach an estimated $164 billion, encompassing rebuilding homes and businesses, cleanup and recovery expenses, and long-term health consequences.[39] Similarly, Canada's economic impacts are also widespread. While the 2023 wildfires did cause significant disruptions in industries such as mining, oil, and gas, particularly in Alberta and British Columbia, they also affected tourism, forestry, agriculture, and other sectors across multiple provinces.[40]

Wildfires in Mexico have traditionally been viewed as a threat to forest ecosystems.[41] However, this perception is evolving as wildfires increasingly impact the wildland-urban interface, posing risks to human populations and infrastructure. In 2024, Mexico recorded over 8000 fires, burning more than 4 million acres across the country.[42] These fires are fueled by dry conditions and high temperatures, often intensified by the El Niño phenomenon.[43] Many fires in Mexico have been linked to agricultural practices, such as slash-and-burn agriculture and burning to improve grazing conditions.[44] Poverty and a lack of forestry culture further exacerbate the problem, as fires are often set to clear land for pasture or to eliminate crop parasites. One study estimates that organic carbon emissions from wildfires are ten times higher than black carbon emissions, contributing substantially to overall carbon emissions in Mexico.[45] Despite the growing threat, Mexico's response to wildfires is challenged by limited resources and a need for more comprehensive fire management strategies.[46] Traditional practices, such as the milpa system, where controlled burning is used to enhance soil fertility, have long been integrated into agricultural and land management approaches.[47] These cultural practices, rooted in traditional knowledge and experience, offer valuable insights for developing sustainable fire management strategies.

---

[38] Wang et al. (2021), p. 252. https://doi.org/10.1038/s41893-020-00645-2.

[39] Li and Yu (2025).

[40] NAT. RES. CAN. *THE STATE OF CANADA'S FORESTS ANNUAL REPORT 2023* (2023). https://www.nrcan.gc.ca/our-natural-resources/forests/state-canadas-forests-report/16496 (accessed September 1, 2024).

[41] Society of Fire Protection Engineers. "Facing the Wildfire Challenge in Europe: Strategies for Risk Reduction and Resilience." *SFPE Europe Digital*, Issue 30 (July 2023). https://www.sfpe.org/publications/periodicals/sfpeeuropedigital/sfpeeurope30/europeissue30feature2 (accessed September 1, 2024).

[42] Coordinación General de Conservación y Restauración. "Cierre Estadístico 2024." Comisión Nacional Forestal, https://www.gob.mx/cms/uploads/attachment/file/965295/Cierre_de_la_Temporada_2024.pdf (accessed February 19, 2025).

[43] Yocom et al. (2010), pp. 1660–1671. https://doi.org/10.1890/09-1209.1.

[44] Martínez Domínguez and Rodríguez Trejo (2008), pp. 709–720.

[45] Corona-Núñez et al. (2020), p. e2020GB006815. https://doi.org/10.1029/2020GB006815.

[46] Rodríguez-Trejo et al. (2011), p. 40. https://doi.org/10.4996/fireecology.0702040.

[47] Nigh and Diemont (2013), pp. e45–e54. https://doi.org/10.1890/120282.

Overall, while California experiences some of the most frequent and costly wildfires in the U.S., the wildfire crisis is multifaceted across the state, nation, and globe, with each region facing unique challenges.[48] Climate change, land management practices, and socio-economic conditions all play critical roles in shaping wildfire dynamics and their impacts worldwide.[49,50,51] In the next section, we will highlight the complexities of state policy structures and programs and distill learnings that can be adopted in other settings.[52,53,54]

4 Rethinking Wildfire Management in California: A Planetary Health Perspective

The field of planetary health has seen the development of various frameworks aimed at enhancing its conceptualization and practical applications. These frameworks provide structured approaches to understanding and addressing the complex interrelationships between human health and the state of the Earth's natural systems. Frameworks such as the Foundational Domains of Planetary Health Education,[55] which offers a comprehensive approach to planetary health education, and the Doughnut Framework, which provides a tool for promoting interdisciplinary policy research in planetary health, are helpful tools for encouraging policy research that are cross-sectoral.[56] This chapter focuses on the Sustainable Development Goals (SDG) Wedding Cake Framework and its application to a new wildfire paradigm in California, viewed through a planetary health lens.[57] This framework offers a distinct approach, clustering the SDGs into three interconnected layers: biosphere, society, and economy. The biosphere layer, the foundation of the model, comprises SDGs related to water, climate, biodiversity, and land use. These goals are fundamental to sustainable development, representing the natural systems upon which all human activity depends.[58] This layer emphasizes the fundamental importance of maintaining Earth's life-support systems for human well-being and development.

---

[48] Williams et al. (2019), p. 892. https://doi.org/10.1029/2018EF001049.

[49] Stephens et al. (2020), p. 354.

[50] Kolden (2019), p. 30. https://doi.org/10.3390/fire2010030.

[51] Abatzoglou and Williams (2016), p. 11770. https://doi.org/10.1073/pnas.1607171113.

[52] Balch et al. (2017), p. 2946. https://doi.org/10.1073/pnas.1617394114.

[53] Schoennagel et al. (2017), p. 4582. https://doi.org/10.1073/pnas.1709437114.

[54] Bowman et al. (2020b), p. 500. https://doi.org/10.1038/s43017-020-0085-3.

[55] Guzmán et al. (2021a, b), pp. e253–e255. https://doi.org/10.1016/S2542-5196(21)00057-5.

[56] Iyer et al. (2021), p. 100844. https://doi.org/10.1016/j.ssmph.2021.100844.

[57] Pham et al. (2023). https://doi.org/10.7189/jogh.13.010305.

[58] Nilsson et al. (2016), pp. 320–322. https://doi.org/10.1038/534320a.

Building upon the biosphere layer is the society layer, encompassing SDGs related to social issues such as poverty, health, education, and gender equality. The interconnectedness of social and environmental goals has been widely recognized, with research highlighting how progress in one area can support advancements in others.[59] This layer acknowledges that social development is intrinsically linked to, and dependent upon, the health of the biosphere.

The economy layer, the uppermost layer of the cake, includes SDGs related to economic growth, industry, and infrastructure. There has been a growing need to reimagine economic systems to align with planetary boundaries and social foundations, echoing the principles of the Doughnut Framework.[60] This layer emphasizes that economic activities should be viewed as a means to achieving social and environmental goals, rather than as ends in themselves.

The SDG Wedding Cake Framework emphasizes the hierarchical nature of these layers, illustrating how the economy and society are embedded within and dependent upon the biosphere. This aligns with the core principles of planetary health, which posits that natural capital cannot be fully substituted by human-made capital.[61] The framework thus provides a clear visual representation of the interdependencies between different aspects of sustainable development.

A key strength of the SDG Wedding Cake Framework is its emphasis on systems thinking in addressing complex planetary health challenges. Such an approach is crucial for understanding the interactions between different SDGs and identifying both synergies and trade-offs in their implementation.[62] This focus on systems thinking aligns with the Systems Thinking and Complexity domain in the Planetary Health Education Framework, highlighting the importance of this skill across diverse conceptualizations of planetary health.[63]

Furthermore, the framework also stresses the importance of interdisciplinary collaboration, requiring cooperation across scientific disciplines, various sectors, and diverse societal actors to effectively address the complex challenges of the Anthropocene.[64] This recognition of the value of diverse perspectives and expertise aligns with the Movement Building and Systems Change domain of the Planetary Health Education Framework.

---

[59] Sachs (2012), pp. 2206–2211. https://doi.org/10.1016/S0140-6736(12)60685-0.

[60] Costanza et al. (2016), pp. 350–355. https://doi.org/10.1016/j.ecolecon.2016.07.001.

[61] Neumayer (1999).

[62] Bai et al. (2016), pp. 351–362. https://doi.org/10.1016/j.gloenvcha.2016.05.008.

[63] Guzmán et al. (2021a, b), pp. e253–e255. https://doi.org/10.1016/S2542-5196(21)00057-5.

The Systems Thinking and Complexity domain in the Planetary Health Education Framework emphasizes understanding the interconnected nature of environmental changes and human health across various spatial and temporal scales. It draws upon ecological approaches to systems thinking, focusing on complex adaptive systems with non-linear relationships, tipping points, and self-organization.

[64] Steffen et al. (2015), p. 1259855. https://doi.org/10.1126/science.1259855.

Building upon the theoretical foundation of the SDG Wedding Cake Framework, the next section transitions to its practical application within the specific context of California wildfires. This exploration will demonstrate how this framework can provide a structured approach to aligning wildfire management strategies with broader sustainability objectives.

5 Applying the SDG Wedding Cake Framework to California's Wildfire Response Strategies

This comprehensive framework in Fig. 1 presents an integrated approach to aligning California's wildfire response strategies with the United Nations Sustainable Development Goals (SDGs). The tables provided cover three interconnected layers: biosphere (see Table 1), society (see Table 2), and economy (see Table 3), demonstrating how wildfire programmatic efforts contribute to broader sustainability objectives across multiple sectors. By linking specific programs and policies to relevant SDGs, we illustrate the multilayered impacts of wildfire management beyond fire prevention and suppression. Each table outlines a menu of potential strategies, example programs, measurable indicators, and recommendations to holistically integrate these efforts with other sustainability goals. This layered approach ensures that our proposed solutions simultaneously support ecosystem health, community resilience, and economic growth. By aligning strategic approaches with the SDGs across these three layers, we can develop comprehensive, long-term solutions while promoting sustainable development in all its dimensions. While the focus of this chapter is on aligning California's strategies with the UN Sustainable Development Goals, it is worth noting that fossil fuel divestment is an important related issue that is not directly addressed in this framework. Divestment from fossil fuels could potentially contribute to long-term wildfire risk reduction by supporting the transition to cleaner energy sources and mitigating climate change impacts. However, this specific strategy falls outside the scope of this chapter. The following example applications of wildfire response strategies in California is meant to be illustrative and not exhaustive.

5.1 *Biosphere Layer*

Building a fire-resilient California requires a proactive and comprehensive approach that prioritizes both immediate mitigation and long-term ecological health. The following biosphere-focused strategies represent potential investment areas towards a sustainable future, aligning wildfire management with the SDGs and promoting a balanced relationship between human activity and natural systems. These strategies

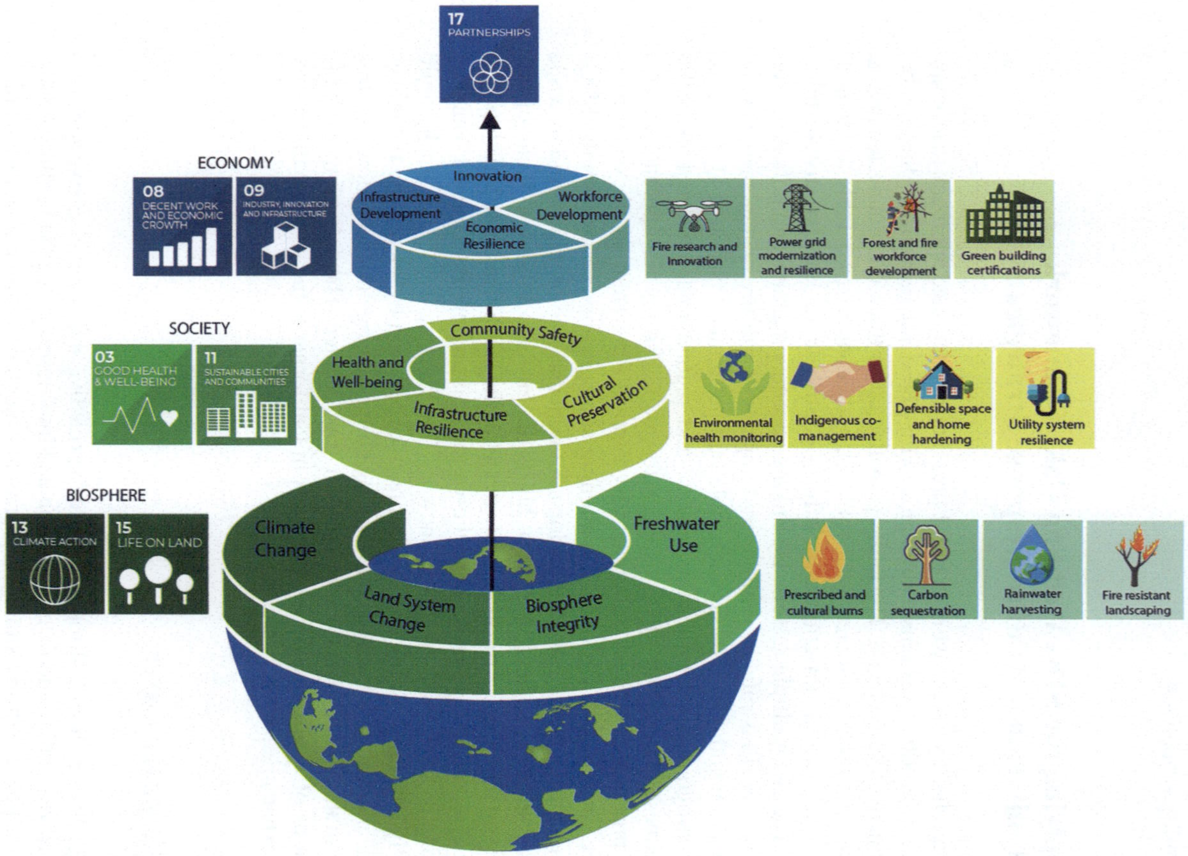

Fig. 1 Graphical representation of the SDG Wedding Cake Framework applied to wildfire management programs in California. This layered approach illustrates how wildfire management strategies align with and contribute to achieving the Sustainable Development Goals (SDGs) across biosphere, society, and economy domains. The figure connects specific strategies and policies (examples shown on the right) with global sustainability goals (examples shown on the left), providing an integrated roadmap for enhancing wildfire resilience and sustainability in California

Table 1 Integrated biosphere strategies for wildfire management and sustainable land stewardship

Biosphere Layer

SDG	Strategic Approach	Example Programs	Key Indicators		
SDG 15: Life on Land	Forest and vegetation management	Forest Stewardship Program	Acres of land treated for fuel reduction	Diversity of native plant species	Impact on wildlife habitat
SDG 15: Life on Land	Controlled burns	Prescribed and cultural burns	Reduction in wildfire intensity	Number of approved burn permits	Return of natural fire cycles

Integrating and aligning with other SDGs

Promote sustainable agriculture (SDG 2)	Support biodiversity conservation (SDG 14)	Create green jobs (SDG 8)	Integrate with carbon sequestration (SDG 13)

Table 2 Community resilience strategies for wildfire adaptation and sustainable infrastructure

Society Layer

SDG	Strategic Approach	Example Programs	Key Indicators		
SDG 11: Sustainable Cities and Communities	Community and infrastructure resistance	Public utility and home hardening programs	Fire-resistant plants, building codes, and zoning regulations	Utility wildfire mitigation plans and vegetation clearance	Indigenous co-management and advisory boards
SDG 11: Sustainable Cities and Communities	Integrated community health response systems	Real-time environmental and health monitoring	Integration of environmental health monitoring and alert systems	Mobile mental health units and clean air spaces	Equitable access to home and health services

Integrating and aligning with other SDGs

Promote native fire-resistant landscaping (SDG 15)	Link with water conservation efforts (SDG 6)	Promote inclusive societies (SDG 16)	Develop wildfire-linked health monitoring systems (SDG 3)

Table 3 Economic strategies for wildfire resilience through workforce development and innovation

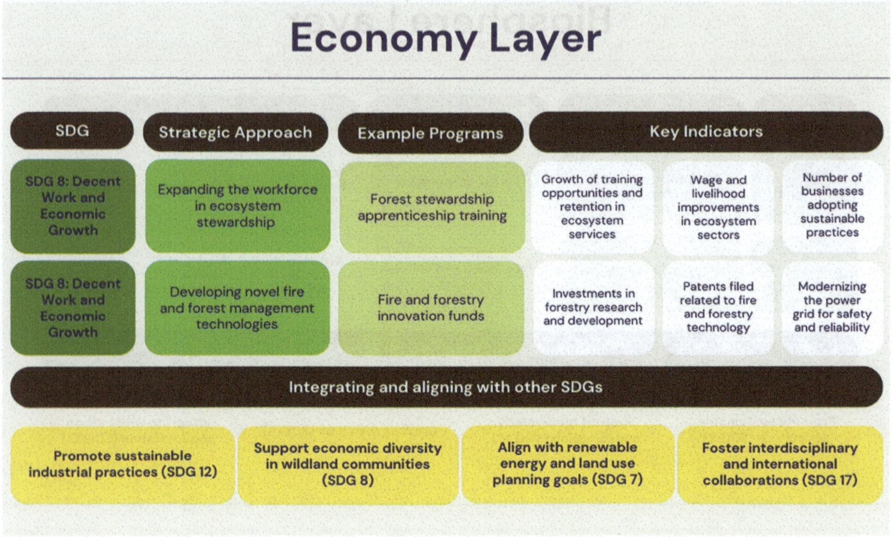

Economy Layer

SDG	Strategic Approach	Example Programs	Key Indicators		
SDG 8: Decent Work and Economic Growth	Expanding the workforce in ecosystem stewardship	Forest stewardship apprenticeship training	Growth of training opportunities and retention in ecosystem services	Wage and livelihood improvements in ecosystem sectors	Number of businesses adopting sustainable practices
SDG 8: Decent Work and Economic Growth	Developing novel fire and forest management technologies	Fire and forestry innovation funds	Investments in forestry research and development	Patents filed related to fire and forestry technology	Modernizing the power grid for safety and reliability

Integrating and aligning with other SDGs

Promote sustainable industrial practices (SDG 12)	Support economic diversity in wildland communities (SDG 8)	Align with renewable energy and land use planning goals (SDG 7)	Foster interdisciplinary and international collaborations (SDG 17)

include expanding the use of prescribed fire, integrating sustainable agriculture, enhancing biodiversity conservation, strengthening carbon sequestration, and improving water conservation.

Expanding the use of prescribed fire[65] **and other controlled burns** like cultural burning[66] is crucial for effective wildfire management and ecosystem health. California, like much of the Western United States, engaged in fire suppression for many decades, which led to an unprecedented build-up of forest fuels and ultimately resulted in extreme wildfire trends.[67] By 2045, California aims to treat up to 1.5 million acres annually with prescribed fire, recognizing its importance in reducing hazardous fuels and minimizing the risk of catastrophic wildfires.[68] Prescribed burns not only protect communities but also improve wildlife habitats, promote

[65] Prescribed burning is the intentional application of fire to vegetation under specific, predetermined conditions to achieve land and resource management objectives.

[66] Cultural burning is a distinct practice from prescribed fire. The Karuk Tribe describes it as a more informal process guided by deep understanding of a specific location, influencing when and how burns are conducted. This practice, carried out by cultural groups, serves numerous purposes, including maintaining travel routes, improving wildlife habitats, attracting animals, managing water resources, controlling pests, caring for culturally significant plants, conservation, and even spiritual practices. Cultural burning is governed by Tribal law and remains under the authority of Tribes, their members, organizations, and cultural fire practitioners.

[67] Zhuang et al. (2024), p. 1736. https://doi.org/10.1038/s41467-024-36832-4.

[68] Administration of Governor Gavin Newsom. *California's Nature-Based Solutions Climate Targets*. 2024. https://resources.ca.gov/-/media/CNRA-Website/Files/Initiatives/Expanding-Nature-Based-Solutions/Californias-NBS-Climate-Targets-2024.pdf.

biodiversity, and restore natural fire cycles in fire-adapted ecosystems.[69] Recent studies have shown that areas treated with prescribed fire experience less severe wildfires, making them easier to control.[70] Moreover, controlled burns can enhance soil health, improve water availability for wildlife, and create defensible space around properties.[71] By combining prescribed fire with other fuel reduction techniques and better respecting and—where supported by members of Indigenous communities—incorporating Indigenous knowledge, land managers can create more resilient landscapes better equipped to withstand the increasing threat of wildfires.[72]

Sustainable agriculture can be integrated into wildfire buffer zones using techniques such as agroforestry. Currently, monoculture farming and traditional buffer zones often lack biodiversity and can be vulnerable to wildfires and other environmental stressors.[73,74] This approach involves combining trees or shrubs with crops or livestock to create more diverse, productive, and resilient land-use systems.[75] In California's fire-prone areas, agroforestry could help reduce wildfire risk while supporting local agriculture.[76] For example, establishing rows of fire-resistant trees or shrubs between agricultural fields and forested areas can create natural firebreaks, protecting crops and communities while providing additional income streams for farmers through timber, fruit, or nut production.[77] Continued research is needed to optimize species selection and management practices for fire resilience and productivity. By promoting multiple functions within fire-prone landscapes, California could enhance food security, support rural livelihoods, and improve ecosystem services.[78]

Enhancing biodiversity conservation could be achieved by expanding wildlife corridors, providing protected pathways allowing animals to move safely between fragmented habitats. Habitat fragmentation due to development and infrastructure

[69] Wade and Lundsford (2025). https://www.fao.org/4/19500e/19500e07.htm.

[70] Wu et al. (2023), p. eadi4123. https://doi.org/10.1126/sciadv.adi4123.

[71] Western Fire Chiefs Association. "The Pros and Cons of Prescribed Burns." WFCA, November 29, 2022. https://wfca.com/wildfire-articles/the-pros-and-cons-of-prescribed-burns/.

[72] United Nations Declaration on the Rights of Indigenous Peoples (UNDRIP). 2007. https://www.un.org/development/desa/indigenouspeoples/declaration-on-the-rights-of-indigenous-peoples.html. While the United States is not a party to the UNDRIP, fire practitioners and policy makers in California (or in other non-party nations) seeking to engage with Tribes may nevertheless find instructive the principles and articles therein relating to respecting Indigenous knowledge, intellectual property, culture, and tradition, and framed by the principle of obtaining free, prior, and informed consent—the right of Indigenous peoples to "maintain, control, protect and develop" the same. (See UNDRIP Articles 31, 5, 11, 12, 18, 19, 25, 26, 27, 29, and 32).

[73] Brimm and Jiménez (2023). https://www.preventionweb.net/news/view/88788.

[74] Abatzoglou and Park Williams (2016), pp. 11770–11775. https://doi.org/10.1073/pnas.1607171113.

[75] Schoeneberger et al. (2017). https://doi.org/10.2737/WO-GTR-96.

[76] Calhoun et al. (2021), pp. 2369–2383. https://doi.org/10.1111/ddi.13374.

[77] Schoeneberger et al. (2017). https://doi.org/10.2737/WO-GTR-96.

[78] *Id.*

poses a significant threat to wildlife populations, isolating them and reducing their ability to thrive.[79] In California's diverse landscape, a statewide network of wildlife corridors could help maintain genetic diversity, support species migration in response to climate change, and reduce human-wildlife conflicts.[80] The economic benefits include reduced costs associated with wildlife-vehicle collisions and increased ecotourism opportunities. Additionally, maintaining connected ecosystems can enhance natural pest control and pollination services, benefiting agricultural productivity.[81] Implementing this program on a larger scale, however, would require collaboration between state agencies, local governments, and private landowners.[82] Incorporating Indigenous knowledge, along with regular monitoring, can inform effective corridor design and management.

Strengthening carbon sequestration involves maintaining healthy forests with balanced carbon cycles, as large, mature trees store significant amounts of carbon and are often more fire-resistant.[83] However, prioritizing timber production in many current forest management strategies over carbon storage can lead to deforestation and degradation, releasing stored carbon and diminishing the forest's capacity to absorb more.[84] Programs like the Family Forest Carbon Program, which provides upfront payments for climate-smart forestry practices, incentivize private landowners to adopt sustainable management, such as extended harvest rotations and reforestation in wildfire-impacted lands, that enhance carbon storage and provide additional income.[85,86] Upfront payments are provided to overcome financial barriers that often prevent landowners from implementing these practices. Such approaches not only increase carbon sequestration but also provide additional income streams for rural landowners, potentially reducing economic pressures to convert forests to other land uses.[87] However, maximizing carbon storage must be balanced with managing fire risk. While carbon storage is important, excessive understory vegetation can increase this risk.[88] Controlled understory management through thinning, prescribed burns, or mechanical treatments can reduce fire risk while maintaining overall forest carbon stocks.

---

[79] Haddad et al. (2015), p. e1500052. https://doi.org/10.1126/sciadv.1500052.

[80] Keeley et al. (2018), pp. 1221–1232. https://doi.org/10.1111/cobi.13095.

[81] Keeley and Syphard (2019), p. 24. https://doi.org/10.1186/s42408-019-0042-6.

[82] Hilty et al. (2020). https://portals.iucn.org/library/sites/library/files/documents/PAG-030-En.pdf.

[83] U.S. Forest Service. "Carbon Management in Fire-Prone Forests" (accessed Sept. 20, 2024).

[84] Woodwell Climate Research Center. "Global Forest Carbon Storage, Explained." https://www.woodwellclimate.org/global-forest-carbon-storage-explained/ (accessed February 19, 2025).

[85] American Forest Foundation. "Family Forest Carbon Program." https://familyforestcarbon.org/ (accessed Sept. 22, 2024).

[86] Charnley et al. (2020), pp. 605–619. https://doi.org/10.1093/jofore/fvaa032.

[87] Chazdon et al. (2020), p. 043002. https://doi.org/10.1088/1748-9326/ab6f8a.

[88] California Wildfire & Forest Resilience Task Force. "Southern California Regional Profile: Carbon Storage." https://wildfiretaskforce.org/southern-california-regional-profile/carbon-storage/ (accessed February 19, 2025).

Improving water conservation could provide alternative water sources during wildfire events and alleviate pressure on municipal water supplies.[89] In many regions, water scarcity is exacerbated by climate change and increasing demand, leading to strain on municipal water supplies, especially during prolonged droughts and wildfire events.[90][91] Implementing rainwater harvesting systems, similar to the system used at the McDonald Observatory in Fort Davis, TX, can capture rainfall from rooftops and store it in underground tanks, providing a sustainable water source for landscaping and household-level emergency fire use.[92] This approach not only conserves precious water resources but also reduces runoff and soil erosion, contributing to overall ecosystem health. Economically, such systems can lead to reduced water bills for property owners and decrease the strain on public water infrastructure.

The biosphere-focused strategies outlined in this section offer a foundation for ecosystem resilience. While other vital biosphere approaches, like water regulation and soil conservation, are not detailed here, they are essential components of a comprehensive wildfire management strategy, and a thorough risk assessment including those elements is crucial for identifying vulnerable areas and prioritizing tailored interventions. By prioritizing sustainable agriculture, carbon sequestration, biodiversity and water conservation, and the return of natural fire intervals to fire-adapted ecosystems, California can make significant progress toward planetary resilience. Building on this foundation of ecosystem resilience, the next section explores societal programs that empower communities to thrive in wildfire-prone landscapes.

5.2 Society Layer

Wildfire resilience in California requires a multifaceted approach that prioritizes equitable access to resources, robust infrastructure, and comprehensive public health protections. This section details key strategies for enhancing community resilience, including ensuring equitable access to resources for home hardening, strategically planning urban development, strengthening utility and public infrastructure, integrating community health response systems, and incorporating Indigenous knowledge and rights into forest management.

---

[89] Campisano et al. (2017), p. 195. https://doi.org/10.1016/j.watres.2017.02.052.

[90] United States Environmental Protection Agency. "Climate Adaptation and Source Water Impacts." https://www.epa.gov/arc-x/climate-adaptation-and-source-water-impacts (accessed February 19, 2025).

[91] United States Geological Survey. "Water Quality After Wildfire." https://www.usgs.gov/mission-areas/water-resources/science/water-quality-after-wildfire (accessed February 19, 2025).

[92] McDonald Observatory Rainwater Harvesting System. Tex. A&M AgriLife Extension. https://rainwaterharvesting.tamu.edu/fire-protection/ (accessed Sep. 24, 2024).

Ensuring equitable access to resources for home and community hardening is paramount for building inclusive wildfire resilience.[93] After all, in the face of wildfire, a home is only as safe as its surrounding community. Expanding programs like the California Wildfire Mitigation Program to provide more comprehensive financial assistance for low- and moderate-income households to retrofit and harden their homes is essential.[94,95] Coupled with this, a statewide fire-resistant building certification program, modeled after the Insurance Institute for Business & Home Safety's Wildfire Prepared Home, can establish clear criteria for fire-resistant construction.[96,97] These criteria could include the use of fire-resistant exterior materials, enclosed eaves, noncombustible decking, and fire-smart landscaping, incorporating native, drought-tolerant plants that support local biodiversity.[98] Drawing on and learning from models from states like Colorado and Oregon, the adoption of robust wildfire-resistant building codes should also be considered.[99,100] This multifaceted approach, incentivizing fire-resistant practices, can lead to reduced property damage, potentially lowering insurance costs for homeowners and businesses, while simultaneously promoting safer, more resilient urban development through the use of sustainable materials.[101] Critically, addressing the equity concerns surrounding the cost of fire-resistant building materials and retrofits is vital to ensure widespread adoption and protect vulnerable populations. Since an individual's home is only as protected as their neighbor's, a community-wide approach to fire safety—where all residents have the means to safeguard their homes—is the most effective way to protect everyone.

Strategic land use planning can directly affect wildfire intensity by influencing development patterns and vegetation management.[102] Many communities today lack comprehensive land use plans that adequately consider wildfire risk, leading to development in high-hazard areas and increased vulnerability.[103] While property-

---

[93] Kramer et al. (2019), p. 641. https://doi.org/10.1071/WF18121.

[94] California Wildfire Mitigation Program (CWMP). Cal. Off. of the State Fire Marshal. https://osfm.fire.ca.gov/committees/california-wildfire-mitigation-program (accessed September 23, 2024).

[95] Barrett and Quarles (2024). https://headwaterseconomics.org/natural-hazards/retrofitting-home-wildfire-resistance/.

[96] Wildfire Prepared Home. "Wildfire Prepared Home Overview." https://wildfireprepared.org/wildfire-prepared-home-overview/ (accessed September 25, 2024).

[97] Green Globes Program. Green Building Initiative. https://www.thegbi.org/green-globes-certification/ (accessed September 24, 2024).

[98] Barrett and Quarles (2024). https://headwaterseconomics.org/wp-content/uploads/2024/06/Wildfire_Retrofit_Report_20240624.pdf.

[99] S.B. 762, 81st Leg., Reg. Sess. (Or. 2021).

[100] S.B. 21-258, 73rd Gen. Assemb., Reg. Sess. (Colo. 2021).

[101] Meacham and McNamee (2023). https://doi.org/10.1007/s10694-023-01490-9.

[102] Headwaters Economies. "Land Use Planning Can Reduce Wildfire Risk to Homes and Communities." https://headwaterseconomics.org/natural-hazards/land-use-planning-wildfire/ (accessed February 11, 2025).

[103] Syphard et al. (2013), p. e71708. https://doi.org/10.1371/journal.pone.0071708.

scale measures are essential, community strategies such as zoning regulations, compact neighborhood development in lower-risk areas, and planned retreats can significantly reduce wildfire risk at the community level.[104] For instance, zoning regulations can guide development away from high-hazard areas and promote fire-resistant construction practices.[105] Communities can also establish overlay zones with specific requirements for vegetation management, building materials, and access, which may supersede existing regulations to mitigate wildfire risk.[106] Importantly, strategic land use planning for wildfire risk reduction should be connected to broader climate change mitigation efforts. By promoting more efficient use of space, reducing reliance on automobiles, and preserving open space, communities can contribute to both wildfire resilience and a more sustainable future.[107]

Enhancing utility and other public infrastructure resilience is crucial not only for preventing utility-caused wildfires, but also for protecting critical infrastructure and minimizing risks to communities and ecosystems in fire-prone areas.[108,109] For instance, the Utility Wildfire Mitigation Program serves as a liaison between key agencies, including CAL FIRE and electric utilities, reviewing annual wildfire mitigation plans and supporting power line inspections.[110,111] To further enhance its effectiveness, the program could incorporate an interagency platform that integrates various existing systems, such as the Wildfire and Landscape Resilience Interagency Tracking System and utility Wildfire Mitigation Plans. Additionally, a publicly accessible tracker could be included to provide comprehensive risk assessments, outline proactive maintenance strategies, and monitor the effectiveness of implemented mitigation measures.[112] This transparent approach would facilitate interagency collaboration of mitigation measures, support a more standardized approach to monitoring and reporting, enhance accountability, inform policy decisions, and empower communities to better prepare for and respond to

---

[104] Natural Hazards Center. "Land Use Adaptations to Wildfire in Unincorporated Communities in Colorado." Natural Hazards Center, University of Colorado Boulder. https://hazards.colorado.edu/quick-response-report/land-use-adaptations-to-wildfire-in-unincorporated-communities-in-colorado (accessed February 13, 2025).

[105] National Association of Counties. "Building Wildfire Resilience." https://www.naco.org/resources/building-wildfire-resilience (accessed February 13, 2025).

[106] Headwaters Economics. "Land Use Planning to Reduce Wildfire Risk." https://headwatereconomics.org/natural-hazards/land-use-planning-wildfire/ (accessed February 7, 2025).

[107] Greenbelt Alliance. "The Critical Role of Greenbelts in Wildfire Resilience." https://greenbelt.org/the-critical-role-of-greenbelts-in-wildfire-resilience/.

[108] Macomber et al. (2024). https://woods.stanford.edu/publications/wildfire-assessing-and-quantifying-risk-exposure-and-mitigation-across-western (accessed February 11, 2025).

[109] Radeloff et al. (2018), p. 3314. https://doi.org/10.1073/pnas.1718850115.

[110] Kramer et al. (2019), p. 641. https://doi.org/10.1071/WF18121.

[111] Syphard and Keeley (2019). https://doi.org/10.3390/fire2040049.

[112] Abatzoglou et al. (2021), p. e2021GL092843. https://doi.org/10.1029/2021GL092843.

wildfire risks.[113,114] Facilitating the sharing of best practices among utilities would further encourage the adoption of innovative fire-resistant infrastructure.[115]

Integrating community health response systems is crucial for addressing the immediate environmental damage from wildfires and controlled burns, as well as proactively mitigating their long-term health effects. Currently, many communities lack adequate resources and infrastructure to effectively monitor and respond to the health impacts of wildfires, leading to delayed or insufficient care for affected populations.[116] Programs like the EPA's Wildfire Smoke Air Monitoring Response Technology, which provides air monitoring equipment to local and Tribal organizations, and the Southern Integrated Prescribed Fire Information System, which integrates weather forecasts, smoke modeling, and air quality data, offer valuable models for enhancing community health monitoring.[117,118,119] Expanding initiatives like these alongside hospital and health services databases, particularly in geographically diverse regions such as California, can provide critical, localized air quality data for more targeted and effective public health responses.[120] This includes establishing networks of permanent monitoring stations in high-risk areas to ensure continuous data for research and public health planning.[121,122] Furthermore, integrating mental health support for wildfire-affected communities is essential to address the long-term psychological impacts of surviving a wildfire. Providing free mental health services, including counseling and trauma-informed therapies, can mitigate the associated mental health burden and reduce associated healthcare costs.[123]

Incorporating Indigenous knowledge and rights into forest management is crucial in preserving cultural and ecological practices, including the practice of

---

[113] Hakes and Caton (2018), p. 15. https://doi.org/10.1016/j.firesaf.2018.04.005.

[114] Cal. Off. of State Fire Marshal. "Utility Wildfire Mitigation." https://osfm.fire.ca.gov/divisions/community-wildfire-preparedness-and-mitigation/wildfire-mitigation-engineering/electric-utility-wildfire-mitigation/ (accessed September 23, 2024).

[115] Coop et al. (2020), p. 659. https://doi.org/10.1093/biosci/biaa063.

[116] Rappold et al. (2017), pp. 6674–6682. https://doi.org/10.1021/acs.est.7b00335.

[117] U.S. Environmental Protection Agency. "Wildfire Smoke Air Monitoring Response Technology (WSMART)." https://www.epa.gov/air-sensor-toolbox/wildfire-smoke-air-monitoring-response--technology-wsmart (accessed Sept. 24, 2024).

[118] Southern Integrated Prescribed Fire Information System (SIPFIS). https://sipc.ce.gatech.edu/SIPFIS/about.php (accessed Sept. 21, 2024).

[119] Hu et al. (2019), p. 1981. https://doi.org/10.3390/ijerph16111981.

[120] Rappold et al. (2019), p. 6906. https://doi.org/10.1021/acs.est.9b01103.

[121] Liu et al. (2015), p. 120. https://doi.org/10.1016/j.envres.2014.10.014.

[122] Zegas et al. (2023). https://woods.stanford.edu/news-events/news/pursuit-clean-air-laying-groundwork-public-school-resilience-wildfire-smoke.

[123] Wildfire Mental Health Collaborative. "Wildfire Mental Health Collaborative in Sonoma County." Sonoma County. https://www.sonomacounty.ca.gov/health/behavioral-health/community-programs/wildfire-mental-health-collaborative (accessed Sept. 24, 2024).

traditional burning.[124] This can be achieved through co-management agreements, like those used by the Karuk Tribe and Parks Canada, or by establishing Indigenous Fire Management Advisory Boards.[125,126,127] These mechanisms enable Indigenous communities to share traditional ecological knowledge and participate in fire management decisions. 'Cultural Fire Keepers' programs further support this by fostering intergenerational knowledge exchange in traditional and modern fire management techniques.[128,129,130] By empowering Indigenous practitioners with a genuine commitment to self-determination and collaboration, California could balance its ecological needs and enhance the cultural relevance of its fire management practices, while addressing historical injustices and promoting more equitable natural resource management.[131,132,133,134] Such steps include authorizing resource agencies to engage with federally recognized tribes to support cultural burning activities without requiring permits for each individual fire.[135]

The strategies outlined in this section—prioritizing equitable access to resources, strategic land use planning, bolstering infrastructure resilience, strengthening community health systems, and supporting cultural fire practices—lay the groundwork for safer, more resilient communities. While other essential programs, such as community-based fire education, emergency response coordination, and early warning systems, are not detailed here, they are crucial components of a comprehensive strategy. A thorough risk and equity analysis will further inform these efforts, ensuring that interventions are strategically targeted and equitably distributed. Building on this second layer, the following section will explore critical economic strategies to support long-term wildfire resilience and minimize the economic burden on communities.

---

[124] Hoffman et al. (2022), pp. 464–481. https://doi.org/10.1139/facets-2021-0072.

[125] Parks Canada. "Indigenous Partners." https://parks.canada.ca/pn-np/ab/banff/culture/autochtones-indigenous (accessed Sept. 22, 2024).

[126] James Bay Advisory Committee on the Environment (JBACE). http://www.jbace.ca/index.php (accessed Sept. 22, 2024).

[127] Pelletier, Flavie. "Fighting Fire with Fire by the Karuk Tribe: Traditional Ecological Knowledge and Wildfire Management in California, USA." *Conservation, Forests and REDD+ Case Studies*. https://www.conservation-strategy.org/en/cs-publications/case-studies/fighting-fire-fire-karuk-tribe-traditional-ecological-knowledge-and (accessed February 10, 2025).

[128] Norgaard and Marks-Block (2019), pp. 519–536. https://doi.org/10.1080/08941920.2018.1540536.

[129] Mistry et al. (2019), p. 20190008. https://doi.org/10.1098/rstb.2019.0008.

[130] Lake et al. (2018), pp. 143–148. https://doi.org/10.1093/jofore/fvy010.

[131] Nikolakis and Roberts (2020), pp. 1–13. https://doi.org/10.1186/s13717-020-00218-4.

[132] Ens et al. (2020), p. 108634. https://doi.org/10.1016/j.biocon.2020.108634.

[133] Reo et al. (2017), pp. 58–68. https://doi.org/10.1177/117718011701300106.

[134] Eriksen and Hankins (2020), p. 32. https://doi.org/10.3390/fire3030032.

[135] S.B. 310, 2023-2024 Reg. Sess. (Cal. 2024).

5.3 Economy Layer

Building long-term economic resilience is crucial to effective wildfire management in California. This section outlines key economic strategies—from diversifying rural economies to fostering interdisciplinary collaborations—that support a holistic approach to wildfire management, aligning with our biosphere, society, and the UN Sustainable Development Goals.

Diversifying the economies of rural and wildland-urban interface communities, while simultaneously enhancing ecosystem restoration, are critical for reducing reliance on single or declining industries and providing long-term sustainable career opportunities. Environmental changes and economic fluctuations threaten industries like agriculture, forestry, and mining, on which many rural economies depend, leading to job insecurity and hindering long-term community resilience.[136] Expanding existing forest stewardship and carbon programs, which provide private landowners access to carbon markets, can prioritize wildfire resilience by offering upfront payments for fire-resistant forest management practices.[137,138,139,140] These practices could include creating fuel breaks, managing understory vegetation, and implementing controlled burns in close collaboration with local fire departments to ensure alignment with community fire protection needs.[141] Furthermore, developing a diverse and robust fire workforce that encompasses ecosystem stewardship objectives beyond fire suppression is equally important.[142,143] Ultimately, a long-term, "whole-of-community" approach is necessary to support wildfire-prone communities not only in managing fire risk but also in achieving economic prosperity.

Promoting sustainable industrial practices can encourage responsible consumption and production while simultaneously reducing wildfire risks. Incentivizing the modernization of the power grid, including investments in sensor technologies that automatically detect and isolate faults, reroute power, and rapidly restore services during wildfire events, is one example.[144,145] This would not only reduce

---

[136] University of Cambridge Institute for Sustainability Leadership and European Trade Union Institute. "Climate Change: Implications for Employment." https://www.cisl.cam.ac.uk/resources/publication-pdfs/climate-change-implications-for-employment.pdf (accessed February 20, 2025).

[137] American Forest Foundation. "Family Forest Carbon Program." https://www.familyforestcarbon.org/ (accessed Sept. 24, 2024).

[138] Steen-Adams et al. (2020), p. 543. https://doi.org/10.1007/s00267-020-01325-3.

[139] Schultz et al. (2019). https://doi.org/10.5751/ES-10940-240321.

[140] Kolden (2019), p. 30. https://doi.org/10.3390/fire2010030.

[141] Roesch-McNally et al. (2018), p. 322. https://doi.org/10.1017/S174217051700041X.

[142] Varney and Jurenci (2024). https://woods.stanford.edu/publications/building-robust-and-representative-wildland-fire-mitigation-and-management-workforce.

[143] Varney and Wallace-Perdomo (2023). https://woods.stanford.edu/publications/developing-and-supporting-federal-wildland-fire-workforce.

[144] Cal. Pub. Utils. Comm'n. *2019 California Smart Grid Annual Report to the Governor and the Legislature* (2020). https://www.cpuc.ca.gov/-/media/cpuc-website/files/uploadedfiles/cpucwebsite/content/about_us/organization/divisions/office_of_governmental_affairs/legislation/2020/2019-smart-grid-annual-report.pdf.

[145] Hurteau et al. (2019), p. 1. https://doi.org/10.1016/j.foreco.2019.05.038.

utility-caused fires but also bolster economic resilience and public safety in vulnerable communities.[146] Additionally, it could promote the integration of distributed energy resources, such as local solar and storage systems, to enhance grid resilience.[147,148] Other supporting programs could include developing sustainable uses for forest biomass in energy production and implementing circular economy principles in post-fire recovery, such as utilizing burned biomass for biochar production or sustainable construction.[149] Developing eco-tourism initiatives that educate visitors about fire ecology and forest management can provide additional economic diversification.[150] Critically, investments in infrastructure resilience must be distributed equitably across all communities, prioritizing those most vulnerable to wildfires.

Innovative financial mechanisms, such as community-based insurance and home hardening insurance discounts, can incentivize accessible insurance coverage, risk reduction, and financial resilience.[151,152] However, a significant problem exists: many homeowners in high wildfire risk areas are priced out or denied coverage by traditional insurance, resulting in underinsurance and financial vulnerability.[153] To address this, insurers can utilize sophisticated catastrophe models to assess and quantify various risks, allowing them to set rates that reflect the level of risk they are undertaking, including offering discounts when risk is reduced. By aggregating risk at the community level, these mechanisms encourage investment in mitigation efforts such as the creation of wildfire risk fuel breaks.[154] Coupled with existing initiatives like insurance discounts for home hardening, these programs can create a multi-layered approach to financial resilience. For instance, California's mandate requiring insurers to offer discounts for individual home hardening measures, such as installing fire-resistant windows or clearing vegetation,[155] and additional discounts for community-level mitigation efforts, like living in a Firewise USA or Fire Risk Reduction Community, can work to enhance both individual and

---

146 Moritz et al. (2020), p. 1107. https://doi.org/10.1073/pnas.1910145117.

147 Higuera and Abatzoglou (2021), p. 1. https://doi.org/10.1111/gcb.15386.

148 Domínguez-García et al. (2022). https://strathprints.strath.ac.uk/79435/1/Moreno_etal_IEEE_PEM_2022_Microgrids_against_wildfires.pdf.

149 Puettmann et al. (2020), p. 119564. https://doi.org/10.1016/j.jclepro.2019.119564.

150 Abrhám et al. (2025), pp. 500–526. https://doi.org/10.15376/biores.20.1.500-526.

151 Marsh McLennan. "Community-Based Catastrophe Insurance." February 2021. https://www.marshmclennan.com/web-assets/insights/publications/2021/february/Community-Based%2D%2DCatastrophe-Insurance-(Final).pdf.

152 Kousky and French (2023), p. 31. https://www.edf.org/sites/default/files/documents/Inclusive%20Insurance%20for%20Climate-Related%20Disasters%20A%20Roadmap%20for%20the%20United%20States.pdf.

153 United Policyholders. "Two-Thirds of Homeowners Underinsured for Wildfire Loss." https://uphelp.org/blog/two-thirds-of-homeowners-underinsured-for-wildfire-loss/ (accessed February 19, 2025).

154 U.S. Department of the Treasury (2022). https://home.treasury.gov/system/files/311/Community-Based%20Catastrophe%20%20Insurance.pdf.

155 Cal. Code Regs. tit. 10, § 2644.9 (2022).

community preparedness.[156] This combination of existing strategies and novel community-based models has the potential to not only reduce individual financial burdens but also foster a culture of shared responsibility in wildfire mitigation.

Fostering interdisciplinary collaborations among diverse stakeholders and rights holders in wildland fire science is crucial for driving innovation in fire and ecosystem management.[157] Establishing international bridges and regional hubs that connect government agencies, Indigenous communities, academic institutions, technology companies, and community members from fire-prone regions can facilitate the co-development of innovative wildfire management solutions.[158] The National Science Foundation's programs to catalyze innovative wildland fire science through diverse collaborations offer a valuable model.[159] However, such initiatives often face challenges in promoting interagency and public data sharing, which is essential for integrating forest science with cutting-edge technological tools.[160] Future programs would benefit from engaging local communities in data collection and monitoring, fostering a sense of stewardship and improving early detection of fire risks. This collaborative approach will not only strengthen the state's wildfire response capabilities but also ensure that diverse perspectives and knowledge systems inform wildfire management strategies.[161]

These economic strategies provide a pathway toward a more robust and sustainable economy in California, particularly within wildfire-prone areas. By prioritizing economic diversification, promoting sustainable industrial practices, incorporating innovative financial mechanisms, and fostering interdisciplinary collaborations, the state can make significant progress toward its wildfire management and economic development goals. While this section does not explore the full range of economic tools available, such as state insurance programs and risk reduction bonds, these are nevertheless important components of a comprehensive wildfire management strategy. A thorough cost-effectiveness analysis, considering the economic vulnerabilities of different regions, is essential for identifying the most impactful interventions.

6 California's Wildfire Challenge: A Path Forward

This chapter illuminates the stark reality of escalating wildfire risks, not only in California, but globally. Our planetary wildfire crisis demands a multifaceted and integrated approach, addressing both immediate threats and the underlying drivers

[156] "Firewise USA," National Fire Protection Association, accessed February 12, 2025, https://www.nfpa.org/education-and-research/wildfire/firewise-usa.

[157] Schoennagel et al. (2017), p. 4582. https://doi.org/10.1073/pnas.1709437114.

[158] Keeley and Syphard (2019), p. 24. https://doi.org/10.1186/s42408-019-0042-6.

[159] National Science Foundation. "Planning Proposals to Catalyze Innovative and Inclusive Wildland Fire Science through Diverse Collaborations." https://www.nsf.gov/pubs/2020/nsf20504/nsf20504.htm (accessed Sept. 21, 2024).

[160] Stephens et al. (2020), p. 354. https://doi.org/10.1002/fee.2181.

[161] Pausas and Keeley (2019), p. 289. https://doi.org/10.1002/fee.2029.

of climate change. Through a case study of California, this chapter explored the crisis using the lens of planetary health, emphasizing the interconnectedness of human well-being and Earth's natural systems. The confluence of climate change, historical land management practices, and evolving fire regimes in California underscores the urgent need for a paradigm shift in wildfire management. California's approach—a blend of legislation, policy frameworks, and traditional ecological knowledge—offers valuable insights for other regions. However, the persistent occurrence of severe wildfires necessitates continuous refinement and improvement.

California's wildfire management involves a complex network of federal, state, local, utility, private, and Indigenous collaborators. While this multi-jurisdictional approach allows for diverse perspectives, it also presents challenges in coordination and cohesive policy implementation. Future efforts should streamline collaboration and develop more unified frameworks. Aligning fire resilience strategies with broader sustainability objectives, as illustrated by the SDG Wedding Cake Framework, demonstrates how wildfire management can simultaneously support ecosystem health, community resilience, and economic growth while addressing multiple Sustainable Development Goals (SDGs). By considering the biosphere, society, and economy together, policymakers can develop more holistic solutions.

Grounded in the principles of planetary health, this analysis highlights the interconnectedness of environmental and human well-being. It calls for the recognition that the health of our planet directly impacts our own. For California, this necessitates leveraging insights from recent research to inform policy and practice. The state must invest in innovative solutions that integrate traditional ecological knowledge with cutting-edge science. These solutions include strategic fuel reduction, promotion of biodiversity to enhance ecosystem resilience, and fostering community-led initiatives that empower local stewardship. While significant, the challenge is not insurmountable. By acknowledging the complex interplay of factors driving wildfire activity and committing to a holistic approach, California can lead the way in forging a sustainable and resilient future.

References

Abatzoglou JT, Park Williams A (2016) Impact of anthropogenic climate change on wildfire across western US forests. Proc Natl Acad Sci 113(42):11770–11775. https://www.nature.com/articles/s41561-020-0641-y

Abatzoglou JT, Williams AP (2016) Impact of anthropogenic climate change on wildfire across western US forests. Proc Nat Acad Sci 113:11770

Abatzoglou JT et al (2021) Anthropogenic influence on recent severe autumn fire weather in the West Coast of the United States. Geophys Res Lett 48:e2021GL092843

Abrhám J, Soukupova J, Prochazka P (2025) Wildfires and tourism in the Mediterranean: balancing conservation and economic interests. BioResources 20(1):500–526

Administration of Governor Gavin Newsom, California's Nature-Based Solutions Climate Targets, 2024

Akdis CA, Nadeau KC (2022) Human and planetary health on fire. Nat Rev Immunol 22(11):651–652

American Forest Foundation, Family Forest Carbon Program. https://familyforestcarbon.org/ (last visited Sept. 22, 2024)

Andela N, Morton DC, Giglio L, Paugam R, Chen Y, Hantson S, van der Werf GR, Randerson JT (2023) The global fire atlas of individual fire size, duration, speed and direction. Nat Commun Earth Environ 4(1):137

Ayars J et al (2023) The 2020 to 2021 California megafires and their impacts on wildlife habitat. Proc Natl Acad Sci 120(48):e2312909120

Bai X, van der Leeuw S, O'Brien K, Berkhout F, Biermann F, Brondizio ES et al (2016) Plausible and desirable futures in the Anthropocene: a new research agenda. Global Environ Change 39:351–362

Balch JK et al (2017) Human-started wildfires expand the fire niche across the United States. Proc Nat Acad Sci 114:2946

Barrett K, Quarles SL (2024) Retrofitting a home for wildfire resistance, Headwaters Econ. (June 24, 2024). https://headwaterseconomics.org/wp-content/uploads/2024/06/Wildfire_Retrofit_Report_20240624.pdf

Bilbao B, Steil L, Urbieta IR, Anderson L, Pinto C, González ME, Millán A, Falleiro RM, Morici E, Ibarnegaray V, Perez-Salicrup DR (2020) Wildfires. Adaptation to climate change risks in ibero-American countries

Bowman DM, Kolden CA, Abatzoglou JT et al (2020a) Vegetation fires in the Anthropocene. Nat Rev Earth Environ. Published online August 18, 2020. https://doi.org/10.1038/s43017-020-0085-3

Bowman DM et al (2020b) Vegetation fires in the Anthropocene. Nat Rev Earth Environ 1:500

Brimm K, Jiménez NP (2023) Farming through wildfire season: the key role of farmers in building wildfire resilience. PreventionWeb, February 13, 2023. https://www.preventionweb.net/news/farming-through-wildfire-season-key-role-farmers-building-wildfire-resilience

Byrne B, Liu J, Bowman KW et al (2024) Carbon emissions from the 2023 Canadian wildfires. Nature. https://doi.org/10.1038/s41586-024-07878-z

CAL FIRE, 2018 Fire Season. https://www.fire.ca.gov/incidents/2018/ (last visited Sept. 1, 2024)

CAL FIRE, 2020 Fire Season. https://www.fire.ca.gov/incidents/2020/ (last visited Sept. 1, 2024)

CAL FIRE, Top 20 Largest California Wildfires. https://www.fire.ca.gov/media/4jandlhh/top20_acres.pdf (last visited Sept. 1, 2024)

Cal. Code Regs. tit. 10, § 2644.9 (2022)

Cal. Off. of State Fire Marshal, Utility Wildfire Mitigation. https://osfm.fire.ca.gov/what-we-do/community-wildfire-preparedness-and-mitigation/utility-wildfire-mitigation (last visited Sept. 23, 2024)

Cal. Pub. Utils. Comm'n, 2019 California Smart Grid Annual Report to the Governor and the Legislature(2020).https://www.cpuc.ca.gov/-/media/cpuc-website/files/uploadedfiles/cpucwebsite/content/about_us/organization/divisions/office_of_governmental_affairs/legislation/2020/2019-smart-grid-annual-report.pdf[1]

Calhoun KL et al (2021) Spatial overlap of wildfire and biodiversity in California highlights gap in non-conifer fire research and management. Divers Distrib 27:2369–2383

California Wildfire & Forest Resilience Task Force. Southern California Regional Profile: Carbon Storage. Accessed 19 Feb 2025. https://wildfiretaskforce.org/southern-california-regional-profile/carbon-storage/

California Wildfire Mitigation Program (CWMP), Cal. Off. of the State Fire Marshal. https://osfm.fire.ca.gov/committees/california-wildfire-mitigation-program (last visited Sep. 23, 2024)

Campisano et al (2017) Urban rainwater harvesting systems: research, implementation and future perspectives. Water Res 115:195

Chang AY, Barry M, Harrington RA (2021) The need to expand the framework of environmental determinants of cardiovascular health from climate change to planetary health: trial by wildfire. Circulation 143(21):2029–2031

Charnley S et al (2020) Engaging family forest landowners in climate change mitigation: approaches to carbon market opportunities. J Forest 118:605–619

Chazdon RL et al (2020) Fostering natural forest regeneration on former agricultural land through economic and policy interventions. Environ Res Lett 15:043002

Chen H et al (2025) Canadian record-breaking wildfires in 2023 and their impact on US air quality. Atmos Environ 342:120941

Coop JD et al (2020) Wildfire-driven forest conversion in Western North American Landscapes. BioScience 70:659

COPERNICUS ATMOSPHERE MONITORING SERV., Canada produced 23% of the global wildfire carbon emissions for 2023. https://atmosphere.copernicus.eu/copernicus-canada-produced-23-global-wildfire-carbon-emissions-2023 (last visited Sept. 1, 2024)

Corona-Núñez RO, Li F, Campo JE (2020) Fires represent an important source of carbon emissions in Mexico. Glob Biogeochem Cycles 34:e2020GB006815

Costanza R et al (2016) Modelling and measuring sustainable wellbeing in connection with the UN sustainable development goals. Ecol Econ 130:350–355

Cova G et al (2023) The outsized role of California's largest wildfires in changing forest burn patterns and coarsening ecosystem scale. Forest Ecol Manag 528:120620

Domínguez-García JL et al (2022) Microgrids against wildfires: a framework for enhancing resilience and reducing risk. University of Strathclyde, Strathclyde. https://strathprints.strath.ac.uk/79435/1/Moreno_etal_IEEE_PEM_2022_Microgrids_against_wildfires.pdf

Ens et al (2020) Indigenous biocultural knowledge in ecosystem science and management: review and insight from Australia. Biol Conserv 245:108634

Eriksen, Hankins DL (2020) Indigenous fire stewardship: revitalizing cultural burning practices in Australia and North America. Fire 3:32

Family Forest Carbon Program. https://familyforestcarbon.org (last visited Sept. 24, 2024)

Firewise USA, National Fire Protection Association. Accessed 12 Feb 2025. https://www.nfpa.org/education-and-research/wildfire/firewise-usa

Friel S (2023) The planetary health equity hothouse: a research agenda for positive change. J Climate Change Health 10:100195

Green Globes Program, Green Building Initiative. https://www.thegbi.org/green-globes-certification/ (last visited Sep. 24, 2024)

Greenbelt Alliance. The critical role of greenbelts in wildfire resilience. Accessed 12 Feb 2025. https://www.greenbelt.org/research/the-critical-role-of-greenbelts-in-wildfire-resilience/

Guzmán CAF, Aguirre AA, Astle B, Barros E, Bayles B, Chimbari M et al (2021a) A framework to guide planetary health education. Lancet Planet Health 5(5):e253–e255

Guzmán C, Potter T, Aguirre AA, Astle B, Barros E, Bayles B, Chimbari M, El-Abbadi N, Evert J, Hackett F, Howard C (2021b) The planetary health education framework. Planet Health Alliance 5:e253–e255

Haddad NM, Brudvig LA, Clobert J, Davies KF, Gonzalez A, Holt RD, Lovejoy TE et al (2015) Habitat fragmentation and its lasting impact on earth's ecosystems. Sci Adv 1(2):e1500052. https://pmc.ncbi.nlm.nih.gov/articles/PMC4643828/

Hakes RS, Caton SE (2018) The ember storm: understanding the burning of structures in wildland-urban interface fires. Fire Safety J 100:15

Hazard Center, Land Use Adaptations to Wildfire in Unincorporated Communities in Colorado. Natural Hazards Center, University of Colorado Boulder. Accessed 13 Feb 2025. https://hazards.colorado.edu/quick-response-report/land-use-adaptations-to-wildfire-in-unincorporated-communities-in-colorado

Headwaters Economics, Land use planning to reduce wildfire risk. Headwaters Economics. Accessed 7 Feb 2025a. https://headwaterseconomics.org/natural-hazards/land-use-planning-wildfire/

Headwaters Economics. Land use planning can reduce wildfire risk to homes and communities. Accessed 11 Feb 2025b. https://headwaterseconomics.org/natural-hazards/land-use-planning-wildfire/

Higuera PE, Abatzoglou JT (2021) Record-setting climate enabled the extraordinary 2020 Fire Season in the Western United States. Glob Change Biol 27:1

Hilty JA et al (2020) Guidelines for conserving connectivity through ecological networks and corridors

Hoffman KM, Christianson AC, Dickson-Hoyle RE, Copes-Gerbitz K, Nikolakis S, Diabo DA, McLeod R et al (2022) The right to burn: barriers and opportunities for indigenous-led fire stewardship in Canada. FACETS 7:464–481. https://www.facetsjournal.com/doi/10.1139/facets-2021-0062

Hu Y et al (2019) Development of a WebGIS-based analysis tool for human health protection from the impacts of prescribed fire smoke in southeastern USA. Int J Environ Res Pub Health 16:1981
Hurteau MD et al (2019) Climate change, fire management, and ecological services in the southwestern US. Forest Ecol Manag 445:1
Intergovernmental Panel on Climate Change (2022) Climate Change 2022: impacts, adaptation and vulnerability. IPCC sixth assessment report. https://www.ipcc.ch/report/ar6/wg2/
Iyer HS, DeVille NV, Stoddard O, Cole J, Myers SS, Li H et al (2021) Sustaining planetary health through systems thinking: public health's critical role. SSM-Population Health 15:100844
James Bay Advisory Committee on the Environment (JBACE). https://www.ccebj-jbace.ca/en/ (last visited Sept. 22, 2024)
Kasischke ES et al (2010) Alaska's changing fire regime — implications for the vulnerability of its boreal forests. Can J For Res 40:1313
Keeley JE, Syphard AD (2019) Twenty-first century California, USA, wildfires: fuel-dominated vs. wind-dominated fires. Fire Ecol 15:24
Keeley ATH et al (2018) Making habitat connectivity a reality. Conserv Biol 32:1221–1232
Kolden CA (2019) We're not doing enough prescribed fire in the Western United States to mitigate wildfire risk. Fire 2:30
Kousky C, French K (2023) Inclusive insurance for climate-related disasters: a roadmap for the United States. Environmental Defense Fund, January 2023, 31. https://blogs.edf.org/markets/wp-content/blogs.dir/32/files/2023/01/Inclusive-Insurance-Report.pdf#page=31
Kousky et al (2018) Wildfire costs in California: the role of electric utilities. Wharton Risk Mgmt. & Decision Processes Ctr.
Kramer A et al (2019) High wildfire damage in interface communities in California. Int J Wildland Fire 28:641
Lake K et al (2018) Returning fire to the land: celebrating traditional knowledge and fire. J Forest 116:143–148
Laurent C et al (2024) Physicochemical characterization of the particulate matter in New Jersey/New York city area, resulting from the Canadian Quebec Wildfires in June 2023. Environ Sci Technol 58(33):14753–14763
Li Z, Yu W (2025) Economic impact of the Los Angeles Wildfires. UCLA Anderson Forecast, February 4, 2025. https://www.anderson.ucla.edu/about/centers/ucla-anderson-forecast/economic-impact-los-angeles-wildfires
Liu Z, Wimberly MC (2016) Direct and indirect effects of climate change on projected future fire regimes in the western United States. Sci Total Environ 542(Part A):65–75. https://doi.org/10.1016/j.scitotenv.2015.10.093
Liu JC et al (2015) A systematic review of the physical health impacts from non-occupational exposure to wildfire smoke. Environ Res 136:120
Macomber E, Wara M, Mastrandrea M (2024) Wildfire: assessing and quantifying risk exposure and mitigation across western utilities. Stanford Woods Institute for the Environment and Stanford Law School's Environmental and Natural Resources Law & Policy Program, Stanford, CA, May 2024. Accessed 23 Sept 2024. https://woodsinstitute.stanford.edu/system/files/publications/Woods_CEPP_Wildire_White_Paper_FINAL.pdf
Marsh McLennan. Community-Based Catastrophe Insurance. February 2021. https://www.marshmclennan.com/web-assets/insights/publications/2021/february/Community%2D%2DBased%2D%2DCatastrophe%2D%2DInsurance%2D%2D(Final).pdf
Martínez Domínguez R, Rodríguez Trejo DA (2008) Forest fires in Mexico and Central América. In: González-Cabán A (ed) Proceedings of the Second International Symposium on Fire Economics, Planning, and Policy: A Global View. U.S. Department of Agriculture, Forest Service, Pacific Southwest Research Station, Albany, CA, pp 709–720. https://www.fs.usda.gov/psw/publications/documents/psw_gtr208en/psw_gtr208en_709-720_dominquez.pdf
McDonald Observatory Rainwater Harvesting System, Tex. A&M AgriLife Extension. https://rainwaterharvesting.tamu.edu/fire-protection/ (last visited Sep. 24, 2024)
Meacham BJ, McNamee M (2023) Conceptual basis for a sustainable and fire resilient built environment. Fire Tech

Miller RK, Field CB, Mach KJ (2020) Barriers and enablers for prescribed burns for wildfire management in California. Nat Sustain 3(2):101–109
Mistry et al (2019) Indigenous fire management: from local practice to public policy. Phil Trans R Soc B 374:20190008
Moreno JM, LagunaDefior C, Barros V, Calvo Buendía E, Marengo JA, Oswald Spring U (eds) pp 435–496
Moritz MA et al (2020) Wildfires and climate change push low-elevation forests across a critical climate threshold for tree regeneration. Proc Natl Acad Sci 117:1107
NAT. RES. CAN., THE STATE OF CANADA'S FORESTS ANNUAL REPORT 2023 (2023). https://www.nrcan.gc.ca/our-natural-resources/forests/state-canadas-forests-report/16496
NAT'L FORESTRY COMM'N OF MEX. (CONAFOR), REPORTE SEMANAL DE RESULTADOS DE INCENDIOS FORESTALES 2024 (2024)
Nat'l Sci. Found., Planning Proposals to Catalyze Innovative and Inclusive Wildland Fire Science through Diverse Collaborations. https://www.nsf.gov/pubs/2022/nsf22122/nsf22122.jsp (last visited Sept. 21, 2024)
National Association of Counties, "Building Wildfire Resilience," National Association of Counties. Accessed 13 Feb 2025. https://www.naco.org/resources/building-wildfire-resilience
Nauslar NJ et al (2018) The 2017 North Bay and Southern California Fires: a case study. Fire 1:18
Neumayer E (1999) Weak versus strong sustainability. Books
Nigh R, Diemont SAW (2013) The Maya Milpa: fire and the legacy of living soil. Front Ecol Environ 11(S1):e45–e54
Nikolakis W, Roberts E (2020) Indigenous fire management: a conceptual review. Ecol Process 9:1–13
Nilsson M, Griggs D, Visbeck M (2016) Policy: map the interactions between sustainable development goals. Nature 534(7607):320–322
Norgaard M, Marks-Block T (2019) Ecological grief and resilience: indigenous experiences of fire management and climate change. Soc Nat Resour 32:519–536
North M, Collins BM, Stephens S (2012) Using fire to increase the scale, benefits, and future maintenance of fuels treatments. J For 110:392–401
Wildfire Prepared Home Overview, WILDFIRE PREPARED HOME. https://wildfireprepared.org/wildfire-prepared-home-overview/ (last visited Sept. 25, 2024)
Pandey P, Huidobro G, Lopes LF, Ganteaume A, Ascoli D, Colaco C et al (2023) A global outlook on increasing wildfire risk: current policy situation and future pathways. Trees Forests People 14:100431
Parks Canada Indigenous Partners. https://www.pc.gc.ca/en/agence-agency/aa-ia (last visited Sept. 22, 2024)
Pausas JG, Keeley JE (2019) Wildfires as an ecosystem service. Front Ecol Environ 17:289
Pelletier, Flavie. Fighting Fire with Fire by the Karuk Tribe: Traditional Ecological Knowledge and Wildfire Management in California, USA. Conservation, Forests and REDD+: Case Studies. Accessed 10 Feb 2025. https://cases.open.ubc.ca/fighting-fire-with-fire-by-the-karuk-tribe-traditional-ecological-knowledge-and-wildfire-management-in-california-usa/
Pham LT, Kumar P, Dahana WD, Nguyen DH (2023) Promoting global health transdisciplinary research for planetary health: towards achieving the 2030 Agenda for Sustainable Development. J Global Health 13
Puettmann M, Sahoo K, Wilson K, Oneil E (2020) Life cycle assessment of biochar produced from forest residues using portable systems. J Clean Prod 250:119564
Radeloff VC et al (2018) Rapid growth of the US wildland-urban interface raises wildfire risk. Proc Natl Acad Sci 115:3314
Rappold AG, Jeanette Reyes, George Pouliot, Wayne E. Cascio, and David Diaz-Sanchez. Community vulnerability to health impacts of wildland fire smoke exposure. Environ Sci Technol 51(12) (2017): 6674-6682. https://pmc.ncbi.nlm.nih.gov/articles/PMC6372951/
Rappold AG et al (2019) Community vulnerability to health impacts of wildland fire smoke exposure. Environ Sci Technol 53:6906

Reid CE et al (2016) Critical review of health impacts of wildfire smoke exposure. Environ Health Persp 124:1334
Reo NJ et al (2017) Factors that support indigenous involvement in multi-actor environmental stewardship. AlterNative: Int J Indigenous Peoples 13:58–68
Rodríguez-Trejo DA et al (2011) The present status of fire ecology, traditional use of fire, and fire management in Mexico and Central America. Fire Ecol 7:40
Roesch-McNally GE et al (2018) The trouble with cover crops: farmers' experiences with overcoming barriers to adoption. Renew Agric Food Syst 33:322
S.B. 21-258, 73rd Gen. Assemb., Reg. Sess. (Colo. 2021)
S.B. 310, 2023-2024 Reg. Sess. (Cal. 2024)
S.B. 762, 81st Leg., Reg. Sess. (Or. 2021)
Sachs JD (2012) From millennium development goals to sustainable development goals. Lancet 379(9832):2206–2211
Safford HD et al (2022) The 2020 California fire season: A year like no other, a return to the past or a harbinger of the future? Glob Ecol Biogeogr 31:962
Schoeneberger MM et al (2017) Agroforestry: enhancing resiliency in U.S. agricultural landscapes under changing conditions, Gen. Tech. Report WO-96
Schoennagel T et al (2017) Adapt to more wildfire in western North American forests as climate changes. Proc Natl Acad Sci 114:4582
Schultz CA et al (2019) Policy tools to address scale mismatches: insights from U.S. forest governance. Ecol Soc 24:21
Society of Fire Protection Engineers, Facing the Wildfire Challenge in Europe: Strategies for Risk Reduction and Resilience, SFPE Europe Digital, Issue 30 (July 2023). https://www.sfpe.org/publications/periodicals/sfpeeuropedigital/sfpeeurope30/europeissue30feature2
Southern Integrated Prescribed Fire Information System (SIPFIS). https://sipc.ce.gatech.edu/SIPFIS/about.php (last visited Sept. 21, 2024)
State of California (2021) California's wildfire and forest resilience action plan
Steen-Adams MM et al (2020) The promise and pitfalls of forest carbon offset programs: a review of the literature. Environ Manag 66:543
Steffen W, Richardson K, Rockström J, Cornell SE, Fetzer I, Bennett EM et al (2015) Planetary boundaries: guiding human development on a changing planet. Science 347(6223):1259855
Stephens SL et al (2018) Drought, tree mortality, and wildfire in forests adapted to frequent fire. BioScience 68(2):77–88
Stephens SL et al (2020) Fire and climate change: conserving seasonally dry forests is still possible. Front Ecol Environ 18:354
Syphard AD, Keeley JE (2019) Factors associated with structure loss in the 2013–2018 California wildfires. Fire 2:49
Syphard AD, Massada AB, Butsic V, Keeley JE (2013) Land use planning and wildfire: development policies influence future probability of housing loss. PLoS ONE 8(8):e71708. https://journals.plos.org/plosone/article?id=10.1371/journal.pone.0071708
Turco M et al (2023) Anthropogenic climate change impacts exacerbate summer forest fires in California. Proc Natl Acad Sci 120(25):e2213815120
U.S. Department of the Treasury. Community-Based Catastrophe Insurance. Washington, D.C.: U.S. Department of the Treasury (2022). https://home.treasury.gov/system/files/311/Community-Based%20Catastrophe%20%20Insurance.pdf
U.S. Forest Serv., Carbon Management in Fire-Prone Forests. https://research.fs.usda.gov/psw/climatechange/carbon (last visited Sept. 20, 2024)
United Policyholders. Two-Thirds of Homeowners Underinsured for Wildfire Loss. Accessed 19 Feb 2025. https://uphelp.org/two-thirds-of-homeowners-underinsured-for-wildfire-loss/
United States Environmental Protection Agency. Climate Adaptation and Source Water Impacts. Accessed 19 Feb 2025. https://www.epa.gov/arc-x/climate-adaptation-and-source-water-impacts
United States Geological Survey. Water Quality After Wildfire. Accessed 19 Feb 2025. https://www.usgs.gov/mission-areas/water-resources/science/water-quality-after-wildfire

University of Cambridge Institute for Sustainability Leadership and European Trade Union Institute. Climate Change: Implications for Employment. Accessed 20 Feb 2025. https://www.cisl.cam.ac.uk/system/files/documents/IPCC%20AR5_Employment_Briefing_WEB_EN.pdf

Varney AA, Jurenci C (2024) Building a robust and representative wildland fire mitigation and management workforce. Stanford Woods Institute for the Environment, November 2024. https://woods.stanford.edu/sites/woods/files/media/file/woods-wildland-fire-mitigation-v05-web.pdf

Varney AA, Wallace-Perdomo S (2023) Developing and supporting the federal wildland fire workforce. Stanford Woods Institute for the Environment, August 2023. https://woodsinstitute.stanford.edu/system/files/publications/Woods-Wildland_Fire_White_Paper.pdf

Wade DD, Lundsford J (2025) Fire as a forest management tool: prescribed burning in the Southern United States. Food and Agriculture Organization of the United Nations. Accessed 12 Feb 2025. https://www.fao.org/4/t9500e/t9500e07.htm

Wang D et al (2021) Economic footprint of California wildfires in 2018. Nat Sustain 4(3):252–260

Westerling AL (2016) Increasing western US forest wildfire activity: sensitivity to changes in the timing of spring. Philos Trans R Soc B: Biol Sci 371:20150178. https://doi.org/10.1098/rstb.2015.0178

Western Fire Chiefs Association, The Pros and Cons of Prescribed Burns, WFCA, November 29, 2022. https://wfca.com/wildfire-articles/the-pros-and-cons-of-prescribed-burns/

Wildfire Mental Health Collaborative in Sonoma County, Sonoma Cnty. https://www.sonomacounty.ca.gov/health/behavioral-health/community-programs/wildfire-mental-health-collaborative (last visited Sep. 24, 2024)

Wildfire Smoke Air Monitoring Response Technology (WSMART), U.S. Env't Prot. Agency. https://www.epa.gov/air-sensor-toolbox/wildfire-smoke-air-monitoring-response-technology-wsmart (last visited Sep. 24, 2024)

Williams AP et al (2019) Observed impacts of anthropogenic climate change on wildfire in California. Earth's Future 7:892

Woodwell Climate Research Center. Global Forest Carbon Storage, Explained. Accessed 19 Feb 2025. https://www.woodwellclimate.org/global-forest-carbon-storage-explained/

Wu X et al (2023) Low-intensity fires mitigate the risk of high-intensity wildfires in California's forests. Sci Adv 9(1):eadi4123. https://doi.org/10.1126/sciadv.adi4123

Yocom LL, Fulé PZ, Brown PM, Cerano J, Villanueva-Díaz J, Falk DA, Cornejo-Oviedo E (2010) El Niño-Southern oscillation effect on a fire regime in northeastern Mexico has changed over time. Ecology 91(6):1660–1671

Zegas G et al (2023) In pursuit of clean air: laying the groundwork for public school resilience to wildfire smoke. Stanford Woods Institute for the Environment, October 2023

Zhuang Y, Kennedy MC, McKenzie D, Lutz JA, Marshall JD (2024) Fire suppression makes wildfires more severe and accentuates climate-driven increases in fire activity. Nat Commun 15(1):1736. https://doi.org/10.1038/s41467-024-46702-0

Fundamental Human Rights and the Climate Crisis: The Critical Interaction Between Law and Medicine

Andrea Rodgers, Anpotowin Jensen, Vivian Vo, and Kajal Khanna

Abstract Focusing on Indigenous youth, this chapter outlines the impacts of climate change on children's health and their fundamental human rights. This chapter further examines the roles and duties of legal and health professionals in the face of the climate crisis and the critical importance of generating and utilizing scientific evidence to protect children's rights. This chapter concludes with an exploration of the potential remedies available to protect children's health through the implementation of fundamental human rights and avenues to integrate law and medicine to remedy human rights violations.

1 Introduction

Climate change is resulting in devastating harms to children's health, with consequences so severe that it is now implicating several fundamental human rights. This chapter explores the roles of legal and health professionals in the face of the climate crisis and what both can do to protect children's human rights. Medical doctors and lawyers have ethical codes that, for centuries, have guided the performance of their

A. Rodgers
Our Children's Trust, U.S. Strategy, Eugene, OR, USA
e-mail: Andrea@ourchildrenstrust.org

A. Jensen
Stanford School of Engineering, Stanford University, Palo Alto, CA, USA
e-mail: Anpojensen@proton.me

V. Vo
Sean N. Parker Center for Allergy and Asthma Research, Stanford University, Stanford, CA, USA
e-mail: v5vo@health.ucsd.edu

K. Khanna (✉)
Department of Emergency Medicine, Stanford University School of Medicine, Stanford, CA, USA
e-mail: kajalk@stanford.edu

G. Antonelli et al. (eds.), *Planetary Health - Laws, Policies and Science on the 'One Health' Approach*, https://doi.org/10.1007/978-3-031-90621-3_7

professional duties. Under the Hippocratic oath, "the physician pledges to prescribe only beneficial treatments, according to his abilities and judgment; to refrain from causing harm or hurt; and to live an exemplary personal and professional life" (Encyclopedia Britannica 2019). Lawyers have ethical duties to zealously advocate for their client's best interests and "[i]n all professional functions a lawyer should be competent, prompt and diligent," and "should seek improvement of the law, access to the legal system, the administration of justice and the quality of service rendered by the legal profession" (American Bar Association 2020). Both lawyers and doctors must preserve confidential information, be honest, keep their clients and patients informed, and provide them with the best information available when recommending a course of medical treatment or a path to vindicate a client's legal rights. These ethical duties are now of particular relevance when assessing how doctors and lawyers can, and should, respond to the escalating climate crisis. This chapter discusses how climate change harms children's health and impacts children's human rights, with a focus on Indigenous youth, the importance of using the most up to date scientific evidence and medical data on the climate crisis to protect the fundamental rights of children, and some potential remedies available to protect children in both the legal and health contexts.

2 Climate Change Impacts

Since the beginning of the Industrial Revolution, human activity in the form of fossil fuel extraction, production, combustion, and land use activities, coupled with the suppression of Indigenous Peoples' rights, has radically and fundamentally altered our planet and the climate system upon which human life depends. While carbon dioxide (CO_2) concentrations in the atmosphere have naturally fluctuated from between 180-300 parts per million (ppm) for hundreds of thousands of years as humanity evolved and developed (Bereiter et al. 2014), the burning of fossil fuels has caused CO_2 levels to go from about 280 ppm in the 1800s to over 420 ppm (Rubino et al. 2019; NOAA 2024). This is a 50% increase since preindustrial times, with the rate of rise in the last 50 years ~7 times greater than the fastest previous documented natural increase in global CO_2 levels (Wendt et al. 2024).

Because of this rapid buildup of CO_2 in the Earth's atmosphere, more incoming solar radiation is retained by the atmosphere and less energy is released back into space (Hansen et al. 2013; von Schuckmann et al. 2016). This causes an "Earth Energy Imbalance or EEI," which scientists say is the "most critical number defining the prospects for continued global warming and climate change" (von Schuckmann et al. 2020). Over the period from 1971 to 2020, the planet experienced a total heat gain of 381 ± 61 ZJ (10^{21} Joules), amounting to an EEI of 0.48 ± 0.1 W m^{-2} (von Schuckmann et al. 2023). The EEI for the period of 2006 through 2020 (0.76 ± 0.20 W m^{-2}) was ~40% greater than the imbalance from 1971–2020 (von Schuckmann et al. 2023).

This excess energy is equivalent to the energy needed to do MRI scans on each individual on Earth ~484,000 times and is leading to "dramatic changes in ecosystems, biodiversity, population extinctions, coral bleaching and infectious disease, as well as redistribution of habitat," and the inundation of the world's coastal communities from sea level rise as a result of ocean warming and cryosphere loss (von Schuckmann et al. 2020; Hansen et al. 2013). Fossil fuel use is pushing our planet's climate and ecosystems to the brink, with devastating harms on human health, particularly on Indigenous Peoples, for whom climate change exacerbates existing "political marginalization, loss of land and resources, human rights violations, discrimination, and unemployment" (United Nations 2007).

It is well documented that climate change is already adversely affecting the physical and mental health of people around the globe, due to heat waves, floods, droughts, and extreme weather events; infectious and vector-borne diseases; and impacts to air, food, and water quality (Ebi et al. 2018). In all regions, extreme heat events "have resulted in human mortality and morbidity" (Intergovernmental Panel on Climate Change 2022) and these and other related extreme events "will significantly increase ill health and premature death from the near-to long-term (Intergovernmental Panel on Climate Change 2022)". These catastrophes, including a youth mental health crisis resulting from climate change, will disproportionately impact the world's most vulnerable people, including children (Intergovernmental Panel on Climate Change 2022; Thiery et al. 2021; Clayton et al. 2021; Philipsborn and Chan 2018).

Children are uniquely vulnerable to climate change health effects due to their higher respiratory rate, lung growth and development, immature immune system, higher metabolic demands, and immature central nervous system (Pacheco 2020; U.S. Global Change Research Program; Watts et al. 2019; United States Court of Appeals for the Ninth Circuit 2019). Other factors also increase the impacts of climate change on the health of children such as "[t]heir ... higher exposure to air, food, and water per unit body weight; unique behavior patterns; and dependence on caregivers[.]" (Ahdoot et al. 2015). Beyond the impact on the physical and mental health of children, increasing ambient temperature may result in increased adverse community environments. As ambient temperature rises, there is an observed increase in violence, crime, and suicide (Anderson and DeLisi 2011; Anderson 2001). Combined, the scientifically established harms described above implicate a number of fundamental human rights at both the individual and community level.

3 Climate Change: A Focus on Indigenous Youth

The 2022 IPCC report states that climate change exacerbates all existing health disparities and environmental health concerns across the globe. To fully understand these impacts and potential solutions discussed later, it is vital to consider the implications of climate change on specific marginalized populations, especially Indigenous youth. According to the book *Global Indigenous Youth: Through Their*

Eyes, from Columbia University's Institute for the Study of Human Rights, Indigenous children are on the frontlines of experiencing climate change and its health impacts (Institute for the Study of Human Rights n.d.). Because Indigenous children's cultures, languages, food systems, and overall well-being are tied to their lands, they are some of the most impacted globally by the effects of climate change, including but not limited to desertification, wildfires, and droughts in their traditional territories.

In addition to the direct impacts of the climate crisis on Indigenous children, governmental action, such as authorizing illegal mining or fossil fuel extraction on sovereign lands, in violation of legally binding obligations under Treaties signed with sovereign Indigenous nations and the United Nations Declaration of the Rights of Indigenous People (UNDRIP), has resulted in greater harm to the health of Indigenous children (Healy et al. 2022). The violation of land and Treaty rights in the United States of the Lakota and other Indigenous Nations of the Oceti Sakowin, for example, have come in the forms of gold and uranium mining, along with other mineral extraction causing high levels of contamination of rivers and water tables, with devastating impacts to the health of the Lakota and other original Nations of the area. Water sources across the state of South Dakota are contaminated and exceed the EPA maximum contaminant level for mercury and uranium. Additionally, five other toxic metals are near the EPA health advisory levels for children weighing 22 pounds (LaGarry 2020; Wienski 2017). Recent research suggests that loss of rights to land and poor human rights conditions compel Indigenous Peoples to migrate from their traditional territories and this contributes to an increase in climate change impacts on traditional Indigenous territories and the Indigenous Nations who come from them (Farrell et al. 2021).

Another area of great importance the connection between climate change and Indigenous youth is the preservation of Indigenous languages. The relationship between Indigenous languages and climate change, especially in the UN and country mitigation campaigns, is poorly understood. What is clear, however, is that the potential loss of an Indigenous language is detrimental to the health of Indigenous Peoples, the ecosystem they have developed the language in for thousands of years, and thus everyone else on the planet. As Indigenous Peoples lose their languages, they also begin to lose their traditional ecological knowledge which protects the environment and mitigates the impacts of climate change. When Indigenous youth lose their languages, their mental health suffers. One tragic example of this are studies which show that Indigenous youth suicide rates increase with the loss of traditional languages (Hallett et al. 2007). Thus, the preservation and protection of Indigenous languages are an important aspect of any human rights-based approach to maintaining the health of both Indigenous youth and the planet in the face of climate change. In line with Article 13 of the UNDRIP, much more must be done to reverse the impacts of colonization on Indigenous languages due to their vital protective nature towards traditional knowledge that promotes the health of the planet as well as Indigenous rights.

4 Climate Change and Human Rights

Indigenous Peoples and the scientific community have officially and repeatedly warned that climate change-related impacts will adversely affect the health and well-being of children and youth, thereby implicating their fundamental human rights. The United Nations Human Rights Office of the High Commissioner (OHCHR) has found that "climate change threatens the effective enjoyment of a range of human rights including those to life, water and sanitation, food, health, housing, self-determination, culture, and development. States have a human rights obligation to prevent the foreseeable adverse effects of climate change and ensure that those affected by it, particularly those in vulnerable situations, have access to effective remedies and means of adaptation to enjoy lives of human dignity" (United Nations Human Rights 2022). The OHCHR submitted a study on the relationship between climate change and the enjoyment of the rights of children protected under the Convention of the Rights of the Child (CRC) (United Nations Human Rights 2017). This study recognized that children are disproportionately impacted by climate change, acknowledged many of the adverse health effects children are experiencing and called on governments to take steps to protect those most vulnerable to the effects of climate change. "The Committee on the Rights of the Child has recalled that States have an obligation to establish non-judicial and judicial mechanisms to provide access to effective remedies for abuses of children's rights by business enterprises, including as a result of their extraterritorial activities and operations, provided that there is a reasonable link between the State and the conduct concerned." (UN Committee on the Rights of the Child General Comment No. 26, 2023).

More recently, the UN High Commissioner for Human Rights issued an analytical study on the impact of loss and damage from the adverse effects of climate change, reaffirming that the human rights of Indigenous Peoples and children "are disproportionately and intersectionally affected" from the effects of climate change (United Nations Human Rights Council 2024). Specifically, the study found that "[human rights-consistent loss and damage assessments, responses and policies should take into account impacts on disproportionately affected groups, including Indigenous Peoples, women and children, and the effects of historical and structural discrimination. The robust collection of disaggregated data and intersectional, gender-responsive and child-responsive analysis can help to ensure that the diverse experiences of loss and damage beyond marketable interests are effectively addressed."

The right to health is first mentioned in the International Covenant on Economic, Social and Cultural Rights (ICESCR), adopted in 1966, which affirms "the right of everyone to the enjoyment of the highest attainable standard of physical and mental health" (United Nations General Assembly 1966). In 2000, the Committee on Economic, Social and Cultural Rights, which monitors the implementation of the ICESCR, issued General Comment 14 which outlines states' obligations to protect the right to health. The right to health is considered an inclusive right extending not only to timely and appropriately health care but also to the socio-economic determinants which are critical for the right to health, including access to a healthy environment and ability to exercise one's culture. Further, the UN General comments have highlighted the impact that the climate crisis will have on essential goods and services that are critical to ensuring the full realization of the right to health. In addition to the protections listed in ICESCR, Article 24 of the United Nations Declaration on the Rights of Indigenous People states that Indigenous Peoples have an equal right to the enjoyment of the highest attainable standard of physical and mental health, and that states shall take the necessary steps to achieve realization of this right.

State obligations to realize the right to health fall into three categories, namely the obligations to respect, protect and fulfill, which are similar to the duties of care lawyers and doctors hold with respect to their clients and patients. Protection of the right to health requires States to prevent third parties from interfering with the right to health. Fulfilling the right to health obligates States to institute administrative, budgetary, judicial and other policy measures to allow for the full realization of the right to health. The right to health is thus an inclusive right. Moving beyond just the provision of good quality, available, accessible, and acceptable health services, the right requires the protection and promotion of a wide scope of the underlying determinants of health which may be impacted by the climate crisis. This includes an obligation to both promote and protect the rights associated with these determinants of health, such as the right to culture and the promotion of language. It also requires utilizing and advancing the best science in advancing treatment or remedies.

The right to culture under Article 15 of the ICESCR recognizes the right of everyone to "take part in cultural life" and "enjoy the benefits of scientific progress and its applications" (United Nations 2009). In General Comment No. 21, the Committee on Economic, Social and Cultural Rights ("CESCR") recalled that "culture is a broad, inclusive concept encompassing all manifestations of human existence" (United Nations 2009). The right to culture is closely related to freedom of expression, freedom of religion and spiritual dignity, and the right to the full development of the human personality (United Nations 2009). Cultural rights seek to ensure that all persons equally enjoy the benefits of scientific progress, which may enhance their enjoyment of other human rights, including those rights that relate to health and adequate living standards. As noted earlier, when Indigenous Peoples lose their native languages, they also lose traditional knowledge that protects the environment and mitigates the impact of climate change. The protection of Indigenous languages is supported by Article 13 of the UNDRIP. UNDRIP, while not legally binding, may represent the development of international legal norms and principles. Additionally, UNDRIP provides an interpretation of the human rights enshrined in other international human rights instruments as they apply to indigenous peoples.

Indigenous people's cultural rights are at risk because of climate change in numerous ways, particularly insofar as the culture is tied to the natural environment. For example, Sophia S., a youth plaintiff in a climate change case against the government of Canada who is of Mi'kmaq descent and a member of the Metepenagiag First Nation, (Climatechasechart 2019) has explained how fire bans in the New

Brunswick area because of climate change-induced extreme heat and drought, has inhibited her community's ability to use sacred fires for a number of traditional cultural practices. Instead of changing government policies that result in high levels of greenhouse gas (GHG) emissions to mitigate against the effects of climate change, the government is making decisions that impede the ability of Sophia and her community to engage in traditional activities that are vital for their cultural identity.

Michelle Bachelet, the first female President of Chile and High Commissioner for Human Rights, has said that "[o]nly urgent climate action can mitigate or avert disasters that will have huge impacts on all of us, especially our children and grandchildren. We must change course now – it is the only option if we want to safeguard humanity" (United Nations News 2021). Governments, health systems, and legal systems must work cohesively, comprehensively, and swiftly to not only address the aforementioned impacts, but to also ensure they are not compounding the ongoing harm and instead are protecting and promoting citizens' fundamental human rights.

5 The Shared Duty to Protect Children in the Climate Crisis: Practical Requirements

5.1 Duties of Lawyers

As the officers entrusted to represent clients as advisor, advocate, negotiator and evaluator, lawyers have a "special responsibility for the quality of justice," (American Bar Association 2020) and must be cognizant to act and represent clients in a manner which protects the fundamental rights of individuals for whom the government and legal system were established, and the society designed to assure such fundamental rights. A lawyer is charged with respecting the legal system and those who serve it, and "when necessary, to challenge the rectitude of official action" and "uphold legal process" (American Bar Association 2020).

As children and young people increasingly turn to lawyers and the judicial branches of their governments to seek redress and protection of their human rights, it becomes essential that their attorneys and advocates understand and utilize the best evidence available to protect their rights in their arguments and presentations to the court, as it is crucial to fulfilling a lawyer's duty to "zealously assert the client's position under the rules of the adversary system" (Hiser n.d.). In the context of climate change and human rights, that best evidence necessarily encompasses the best available climate science, since it is the climate science which dictates what needs to be done to avert the climate crisis and protect humanity. Dr. Heather Price has equated a climate scientist with doctors for the Earth: "there are doctors for the body, they can specialize in the body when it's young, like a pediatrician, or when it's old, like geriatrics, or someone that specializes in neurology compared to a neurosurgeon – they are both working on the human head but in different ways … . All this just for the diversity of human bodies. I think about climate scientists the same way" (Hiser n.d.).

Climate scientists have indeed created a 'prescription' to stabilize the climate system such that it is one in which humans can survive, which indicates that global CO_2 concentrations in the year 2100 should be no more than 350 parts per million (ppm). "Stabilization of climate, the goal of the universally agreed United Nations Framework Convention on Climate Change (UNFCCC) in 1992 and the Paris Agreement in 2015, requires that EEI be reduced to approximately zero to achieve Earth's system quasi-equilibrium" (von Schuckmann et al. 2020). To do so, the atmospheric concentration of CO_2 must be reduced to approximately 350 ppm by the end of the century "to increase heat radiation to space by [the current level of the imbalance] bringing Earth back toward energy balance" (von Schuckmann et al. 2020). A global GHG emission reduction and sequestration pathway back to 350 ppm by 2100 would stabilize long-term global heating at no more than 1 °C above pre-industrial temperatures, with a short-term peak of approximately 1.3 °C as a global average.

International politically-agreed targets of 1.5 °C or 2 °C, which are referenced in the Paris Agreement are neither science-based nor protective of human rights, particularly for indigenous children. These targets fail to avoid dangerous levels of warming and risk significant and widespread impacts to millions across the globe and the Intergovernmental Panel on Climate Change has recognized these are targets "are not considered 'safe' for most nations, communities, ecosystems and sectors … ." (Intergovernmental Panel on Climate Change (IPCC) 2018). Reducing the atmospheric concentration of CO_2 to below 350 ppm is critical for preserving a safe climate system and central to a human-rights compliant standard for climate mitigation and essential for respecting, protecting and fulfilling fundamental human rights.

Fulfilling a lawyer's duty to protect the fundamental human rights of their clients thus demands that the evidence presented to courts be grounded in the best-available science of achieving an EEI of zero, which scientists believe can only be achieved by returning global CO_2 concentrations to below 350 ppm (Hansen et al. 2008, 2013; Veron et al. 2009; Frieler et al. 2013; von Schuckmann et al. 2020). This standard would provide a clear benchmark to states for the fulfillment of human rights in a climate change-affected world and would satisfy a lawyer's multiple duties to, among other things, zealously advocate for their client and uphold the legal system.

5.2 *Duties of Doctors*

In addition to the provision of essential health services and goods, protecting children's human rights in the face of a climate emergency requires a national health system to have a means of community participation in priority setting as well as transparent accountability mechanisms. It is critical that we integrate best practices into the healthcare system for identifying, treating, and providing care to children harmed by the climate crisis. To support legal action to protect health rights, lawyers, physicians, advocates and community members need to work together to identify and collect evidence that supports constitutional obligations regarding climate and health. For example, the identified mental health impacts of the climate crisis highlight the need to expand current screening tools used to identify adverse childhood experiences to include alternative sources of toxic stress. Data gathered from the implementation of this modified screening tool could provide evidence for setting environmental standards for the protection of pediatric mental health.

National Healthcare systems need to work with Sovereign Indigenous Nations to both identify climate-related violations of human rights related to health as well as best practices to protect and promote fundamental rights of Indigenous Peoples. While states have some discretion in setting environmental standards for the protection of human rights, such standards should be "consistent with all relevant international environmental, health and safety standards," and "should take into account the best available science" (United Nations Human Rights n.d.). Doctors play a critical role in providing evidence that can shape decisions around appropriate health standards.

6 Integrating Law and Medicine to Remedy Human Rights Violations from the Climate Crisis

Given the climate emergency, children are turning to their medical and legal systems to safeguard their human rights. In 2015, twenty-one young Americans, including eleven Black, Brown, and Indigenous youth, filed a constitutional climate lawsuit, *Juliana v. United States*, seeking to protect their fundamental rights. This case asserts that, through the government's past and ongoing affirmative actions that cause climate change, it has violated the youngest generation's constitutional rights to life, liberty, property, and equal protection of the law, as well as failed to protect essential public trust resources. In this litigation, supported by top experts including pediatricians and children's mental health experts, federal courts have affirmed "that the federal government has long promoted fossil fuel use despite knowing that it can cause catastrophic climate change" (US Court of Appeals 2020) and "has long understood the risks of fossil fuel use and increasing carbon dioxide emissions" (US Court of Appeals 2020). While the 9th Circuit Court of Appeals ordered the case be dismissed for lack of standing, the youth continue in their efforts to get the case back on the track to trial. Children have filed similar cases in courts around the world seeking to protect their fundamental rights, many of which have been successful. For example, in August 2023 sixteen youth in Montana secured an unprecedented legal victory against their state government, with a court finding that the state's fossil fuel-based energy system violated their fundamental rights preserved in Montana's constitution. The court found that "[u]ntil atmospheric GHG concentrations are reduced, extreme weather events and other climactic events such as droughts and heatwaves will occur more frequently and in greater magnitude, and

Plaintiffs will be unable to live clean and healthy lives in Montana." Based on the scientific testimony presented, the court acknowledged that "[c]hildren are uniquely vulnerable to the consequences of climate change, which harms their physical and psychological health and safety, interferes with family and cultural foundations and integrity, and causes economic deprivations" and "[b]ecause of their unique vulnerabilities, their stages of development as youth, and their average longevity on the planet in the future, Plaintiffs face lifelong hardships resulting from climate change." (Held v. State of Montana, 2023). In June 2024, thirteen youth from Hawai'i, including several Native Hawaiian youth, settled their constitutional climate case against their state government, with the state agreeing to, among other things, achieve zero emissions in transportation no later than 2045 (Navahine v. State of Hawaii, 2024). In the agreement, the parties agreed that "the best scientific evidence today shows that correcting Earth's energy imbalance requires reducing atmospheric carbon dioxide to less than 350 parts per million this century."

As doctors are on the forefront of providing medical care to children impacted by the climate crisis, they too can play a critical role in the identification and documentation of the violation of the right to life, the right to health, right to culture and other fundamental human rights implicated in the climate crisis. Lawyers and courts can and should rely on the expertise of medical doctors in assessing what is needed to respect, protect, and fulfill human rights. The testimony of medical professionals in both the *Held* and *Navahine* cases proved central to the youth obtaining remedies designed to protect their fundamental rights. Both doctors and lawyers have an obligation to take urgent action on the progressive realization of the right to health and to immediately establish a minimal essential level of the right to health. Doctors, lawyers, climate scientists and community members need to work together to identify the core aspects of climate crisis mitigation that form part of the rights to life, health, and culture.

7 Conclusion

This chapter highlights the intersection between law, health and science and the responsibilities that individual practitioners have in using scientific evidence to advocate for change. Focusing on Indigenous youth, we describe the health harms children are experiencing due to the climate crisis and outline potential remedies using human rights, specifically the right to health, life, culture and other rights contained in treaties and the UNDRIP. A human-rights based approach to climate crisis mitigation is fundamental to addressing the physical and mental health harms from climate change. To successfully respect, fulfill and protect universal human rights, we must take an interdisciplinary participatory approach in gathering the most current scientific evidence and medical data which can identify violations, underscore obligations, and propose solutions resulting in harm minimization from the climate crisis.

References

Ahdoot S, Pacheco SE, The Council on Environmental Health (2015) Global Climate Change and Children's Health. Available at: https://pediatrics.aappublications.org/content/pediatrics/136/5/e1468.full.pdf. Accessed 8 June 2022

American Bar Association (2020) Model rules of professional conduct. Available at: https://www.americanbar.org/groups/professional_responsibility/publications/model_rules_of_professional_conduct/. Accessed 7 June 2022

Anderson C (2001) Heat and violence. Curr Dir Psychol Sci 10:33–38. https://doi.org/10.1111/1467-8721.00109

Anderson CA, Delisi M (2011) Implications of global climate change for violence in developed and developing countries, edition. na

Bereiter B, Fischer H, Schwander J, Stocker TF (2014) Diffusive equilibration of N_2, O_2 and CO_2 mixing ratios in a 1.5-million-years-old ice core. Cryosphere 8:245–256. https://doi.org/10.5194/tc-8-245-2014

Clayton S, Manning CM, Speiser M, Hill AN (2021) Mental health and our changing climate: impacts, inequities, responses. American Psychological Association, and ecoAmerica, Washington, D.C.

Climatechasechart (2019) La Rose v. Her Majesty the Queen. Available at: http://climatecasechart.com/non-us-case/la-rose-v-her-majesty-the-queen/. Accessed 8 June 2022

Ebi KL, Balbus JM, Luber G, Bole A, Crimmins A, Glass G, Saha S, Shimamoto MM, Trtanj J, White-Newsome JL (2018) Human health. In: Impacts, risks, and adaptation in the United States: Fourth National Climate Assessment, Volume II edition. Washington, DC, USA

Encyclopedia Britannica (2019) Hippocratic Oath. Available at: https://www.britannica.com/topic/Hippocratic-oath. Accessed 7 June 2022

Farrell J, Burow PB, Mcconnell K, Bayham J, Whyte K et al (2021) Effects of land dispossession and forced migration on Indigenous peoples in North America. Science 374:eabe4943. https://doi.org/10.1126/science.abe4943

First Circuit Court of the State of Hawai'i, *Navahine F., a Minor v. Dept. of Transportation et al.*, Civ. No. 1CCV-22-0000631, ruling of 6 April 2023

Frieler K, Meinshausen M, Golly A, Mengel M, Lebek K et al (2013) Limiting global warming to 2 °C is unlikely to save most coral reefs. Nat Climate Change 3:165–170. https://doi.org/10.1038/nclimate1674

Hallett D, Chandler MJ, Lalonde CE (2007) Aboriginal language knowledge and youth suicide. Cogn Dev 22:392–399. https://doi.org/10.1016/j.cogdev.2007.02.001

Hansen J, Masson-Delmotte V, Pagani M, Raymo M, Royer DL et al (2008) Target atmospheric CO2: where should humanity aim? Open Atmos Sci J 2:217–231

Hansen J, Kharecha P, Sato M, Masson-Delmotte V, Ackerman F et al (2013) Assessing "dangerous climate change": required reduction of carbon emissions to protect young people, future generations and nature. PLoS One 8:e81648. https://doi.org/10.1371/journal.pone.0081648

Healy JP, Jensen A, Power MB, Mckibben B, Cohen G et al (2022) COVID-19 and climate change: crises of structural racism. J Clim Change Health 5:100092. https://doi.org/10.1016/j.joclim.2021.100092

Hiser K (n.d.) "Hey, Got a Light?" How chemistry, law, and climate justice are related. Available at: https://medium.com/field-notes-teaching-climate-change-in-higher/hey-got-a-light-how-chemistry-law-and-climate-justice-are-related-61f5473abd69. Accessed 8 June 2022

Institute for the Study of Human Rights (n.d.) Global indigenous youth: through their eyes. Available at: https://academiccommons.columbia.edu/doi/10.7916/d8-dh2w-rz29. Accessed 8 June 2022

Intergovernmental Panel on Climate Change (2022) Climate change 2022. Impacts, adaptation and vulnerability. Available at: https://www.ipcc.ch/report/ar6/wg2/downloads/report/IPCC_AR6_WGII_SummaryForPolicymakers.pdf. Accessed 7 June 2022

Intergovernmental Panel on Climate Change (IPCC) (2018) Global warming of 1.5 °C. Available at: https://www.ipcc.ch/sr15/. Accessed 9 June 2022

Lagarry HE (2020) Expert opinion on the potential contamination of surface waters in western south Dakota by unconfined ISL uranium mine Lixiviant and historical open pit mining. Available at: https://lakota-prod.s3-us-west-2.amazonaws.com/uploads/LaGarry-Report.pdf. Accessed 8 June 2022

No. CDV-2020-307 (Mont. Dist. Ct. Aug. 14, 2023)

NOAA, NOAA GML Data, https://gml.noaa.gov/webdata/ccgg/trends/co2/co2_annmean_mlo.txt. Accessed 17 Dec 2024

Pacheco SE (2020) Catastrophic effects of climate change on children's health start before birth. J Clin Invest 130:562–564. https://doi.org/10.1172/jci135005

Philipsborn RP, Chan K (2018) Climate change and global child health. Pediatrics 141. https://doi.org/10.1542/peds.2017-3774

Rubino M, Etheridge D, Thornton D, Howden R, Allison C, Francey R, Langenfelds R, Steele LP, Trudinger C, Spencer DA, Curran MAJ, van Ommen T, Smith AM (2019) Revised records of atmospheric trace gases CO2, CH4, N2O, and δ13C-CO2 over the last 2000 years from Law Dome, Antarctica. Earth Syst Sci Data 11:473–492. https://doi.org/10.5194/essd-11-473-2019

Thiery W, Lange S, Rogelj J, Schleussner CF, Gudmundsson L et al (2021) Intergenerational inequities in exposure to climate extremes. Science 374:158–160. https://doi.org/10.1126/science.abi7339

U.S. Global Change Research Program. Fourth National Climate Assessment. Chapter 24: Northwest. Available at: https://nca2018.globalchange.gov/chapter/24/. Accessed 8 June 2022

United Nations (2007) Department of Economic and Social Affairs: Indigenous Peoples - Climate Change. Available at: https://www.un.org/development/desa/indigenouspeoples/climate-change.html#:~:text=The%20effects%20of%20climate%20change%20on%20indigenous%20peoples&text=Climate%20change%20exacerbates%20the%20difficulties,rights%20violations%2C%20discrimination%20and%20unemployment. Accessed 7 June 2022

United Nations (2009) Economic and Social Council. Available at: https://www.refworld.org/docid/4ed35bae2.html. Accessed 8 June 2022

United Nations General Assembly (1966) International Covenant on Economic, Social and Cultural Rights. United Nations, Treaty Series, vol 993, p 3. Available at: https://www.refworld.org/docid/3ae6b36c0.html. Accessed 8 June 2022

United Nations Human Rights (2017) Climate change and the full and effective enjoyment of the rights of the child. Available at: https://www.ohchr.org/sites/default/files/Documents/Issues/ClimateChange/RightsChild/ChilrenOnePager.pdf. Accessed 8 June 2022

United Nations Human Rights (2022) OHCHR and climate change. Available at: https://www.ohchr.org/en/climate-change#:~:text=Climate%20change%20threatens%20the%20effective,%2Ddetermination%2C%20culture%20and%20development. Accessed 8 June 2022

United Nations Human Rights (n.d.) Special Rapporteur on human rights and the environment. Available at: https://www.ohchr.org/en/special-procedures/sr-environment. Accessed 8 June 2022

United Nations News (2021) Human Rights chief says climate action only way to 'safeguard humanity' Available at: https://news.un.org/en/story/2021/10/1104352. Accessed 8 June 2022

United States Court of Appeals for the Ninth Circuit (2019) Brief of amici curiae public health experts, public health organizations, and doctors in support of plaintiffs-appellees seeking affirmance. Available at: https://news.thoracic.org/washington-letter/2019/resources/public-health-experts-brief.pdf. Accessed 7 June 2022

US Court of Appeals (2020) Juliana v. United States. Available at: https://cdn.ca9.uscourts.gov/datastore/opinions/2020/01/17/18-36082.pdf. Accessed 8 June 2022

Veron JEN, Hoegh-Guldberg O, Lenton TM, Lough JM, Obura DO et al (2009) The coral reef crisis: the critical importance of<350ppm CO2. Mar Pollut Bull 58:1428–1436. https://doi.org/10.1016/j.marpolbul.2009.09.009

von Schuckmann K, Palmer M, Trenberth K et al (2016) An imperative to monitor Earth's energy imbalance. Nat Climate Change 6:138–144. https://doi.org/10.1038/nclimate2876

von Schuckmann K, Cheng L, Palmer MD, Hansen J, Tassone C et al (2020) Heat stored in the Earth system: where does the energy go? Earth Syst Sci Data 12:2013–2041. https://doi.org/10.5194/essd-12-2013-2020

von Schuckmann K, Minière A, Gues F, Cuesta-Valero FJ, Kirchengast G, Adusumilli S, Straneo F, Ablain M, Allan RP, Barker PM, Beltrami H, Blazquez A, Boyer T, Cheng L, Church J, Desbruyeres D, Dolman H, Domingues CM, García-García A, Giglio D, Gilson JE, Gorfer M, Haimberger L, Hakuba MZ, Hendricks S, Hosoda S, Johnson GC, Killick R, King B, Kolodziejczyk N, Korosov A, Krinner G, Kuusela M, Landerer FW, Langer M, Lavergne T, Lawrence I, Li Y, Lyman J, Marti F, Marzeion B, Mayer M, MacDougall AH, McDougall T, Monselesan DP, Nitzbon J, Otosaka I, Peng J, Purkey S, Roemmich D, Sato K, Sato K, Savita A, Schweiger A, Shepherd A, Seneviratne SI, Simons L, Slater DA, Slater T, Steiner AK, Suga T, Szekely T, Thiery W, Timmermans M-L, Vanderkelen I, Wjiffels SE, Wu T, Zemp M (2023) Heat stored in the Earth system 1960–2020: where does the energy go? Earth Syst Sci Data 15:1675–1709. https://doi.org/10.5194/essd-15-1675-2023

Watts N, Amann M, Arnell N, Ayeb-Karlsson S, Belesova K et al (2019) The 2019 report of The Lancet Countdown on health and climate change: ensuring that the health of a child born today is not defined by a changing climate. Lancet 394:1836–1878. https://doi.org/10.1016/s0140-6736(19)32596-6

Wendt KA et al (2024) Southern Ocean drives multidecadal atmospheric CO2 rise during Heinrich Stadials. Proc Natl Acad Sci 121(21):e2319652121

Wienski K (2017) Leading Health Challenges Pine Ridge Reservation, South Dakota Oglala Lakota Sioux. Available at: https://juniperpublishers.com/jojph/pdf/JOJPH.MS.ID.555574.pdf. Accessed 8 June 2022

One Health: Oriented Directions to Bring Science to Policy in Climate Change Mitigation and Adaptation

Skye S. Flanigan, Sweta Waghela, Vanitha Sampath, and Kari C. Nadeau

Abstract Increased anthropogenic activity since the beginning of industrialization has led to rapid changes in the human exposome, which is the measure of all the exposures of an individual in a lifetime and how those exposures relate to health. These anthropogenic activities such as increased burning of fossil fuels, deforestation, loss of biodiversity, urbanization, and modern lifestyles and agricultural practices are having detrimental consequences not only on human health, but also on animal and planetary health as these are inextricably linked and interdependent. This has led to increased consideration of One Health approaches, which aims to sustainably balance and optimize the health of people, animals and ecosystems. The first major policy aimed towards mitigating the health effects of environmental pollutants was the 1963 Clean Air Act enacted by the United States. Nearly a decade later in 1972, the First Earth Summit, was held in Sweden. In 2007, in a conference on Avian and Pandemic Influenza in India, representatives of 111 countries and 29 international organizations were encouraged to develop the One Health concept by building linkages between human and animal health systems for pandemic preparedness and human security. The following year, a strategic framework for reducing risks of infectious diseases at the animal–human–ecosystems Interface was published. In 2015, 196 countries adopted the Paris Agreement, a legally binding international treaty on climate change, that addresses mitigation, adaptation, and finance. It aims for countries to limit global temperature rise this century to well

S. S. Flanigan · S. Waghela
Center for Climate, Health, and the Global Environment, Harvard T.H. Chan School of Public Health, Boston, MA, USA
e-mail: flanigan@hsph.harvard.edu; swaghela@hsph.harvard.edu

V. Sampath · K. C. Nadeau (✉)
Department of Environmental Health, Harvard T. H. Chan School of Public Health, Boston, MA, USA
e-mail: vsampath@hsph.harvard.edu; knadeau@hsph.harvard.edu

G. Antonelli et al. (eds.), *Planetary Health - Laws, Policies and Science on the 'One Health' Approach*, https://doi.org/10.1007/978-3-031-90621-3_8

below 2°Celsius above pre-industrial levels, while pursuing efforts to limit the increase to 1.5°Celsius by 2100. However, current consensus is that we are falling short of this goal. Increased efforts are needed on a global front. Renewed efforts are needed to meet the goals of the Paris Agreement and safeguard human health from climate change.

1 Introduction

The field of human health has traditionally focused on individual and public health, with an emphasis on treating diseases and preventing illness through medical and public health interventions. However, it is increasingly recognized that this approach is limited, and that human health needs to be considered in the context of the environment that it is integrally connected to. This has become clear as humans are rapidly changing the environment through increased greenhouse gas emissions (GHG) released from burning fossil fuels; depletion of natural resources; loss of biodiversity; and an increase in emissions of synthetic pollutants, many of which are resistant to degradation and are rapidly accumulating in our environment, such as plastics.

Increased anthropogenic activity since the Industrial Age has led to adverse planetary health effects such as increasing global temperatures; air, water, and soil pollution; and frequency and intensity of extreme weather effects due to climate change. They are also endangering the components of good health, such as access to clean air, safe drinking water, and nutritious foods, while also increasing the risk of adverse health effects, such as cardiovascular and respiratory illnesses and vector-borne diseases, such as malaria and dengue fever.[1] These exposures, collectively contributing to the human exposome, highlight how human health is profoundly shaped by environmental factors.

With the increased recognition that human health is connected to planetary health, there is a push for a One Health approach, which recognizes that threats to human health can only be addressed through an integrated, unified approach that aims to sustainably balance and optimize the health of people, animals, plants, and planetary ecosystems.[2] The 2024 Lancet report on health and climate change emphasizes that public health strategies must now incorporate Planetary Health and One Health principles to build sustainable health systems. The report underscores the need for traditional health systems to evolve to address global challenges such as pandemics, food insecurity, and the health impacts of climate change. Incorporating ecological and animal health perspectives is essential for building resilient health systems capable of addressing multifaceted impacts of climate change.[3]

---

[1] Myers (2017), pp. 2860–2868.

[2] CDC (2024a).

[3] Vora et al. (2024).

The One Health approach is critical for addressing complex challenges like zoonotic diseases, which are often driven by human-induced environmental changes, such as deforestation, urbanization, and the intensification of agriculture.[4] The COVID-19 pandemic is a critical example of how human encroachment into wildlife habitats can increase the risk of zoonotic spillover by increasing human-wildlife interactions and facilitating the transmission of pathogens from animals to humans. Air pollution and climate change increase incidence and exacerbation of respiratory diseases[5] necessitating the need for Planetary Health approaches. Integrating these approaches into policy and practice is crucial for fostering a more resilient and sustainable global health system, ensuring the well-being of current and future generations.[6]

In this chapter, we discuss the effects of anthropogenic activity on planetary human health, existing health inequities, climate change, One Health approaches, and policies to advance climate change adaptation and mitigation while prioritizing environmental justice. We also suggest ways to advance the understanding of One Health principles.

2 Anthropogenic Activity and the Adverse Effects on Planetary and Human Health

Human activities such as burning fossil fuels, deforestation, urbanization, and unsustainable agricultural practices have led to ecological disruptions, land degradation, and biodiversity loss. Historically, farming practices were the first major disturbance to natural ecosystems, but since the Industrial Revolution, the increase in manufacturing, logging, and mining for fossil fuels and other precious materials has amplified the ecological impacts on the natural environment. These changes have led to increases in vector-borne diseases, more frequent and intense extreme weather events, rising temperatures, and increases in toxic pollutants in our air, soil, and water. Plant and animal species are also threatened by human activities that destroy habitats, pollute the environment, and cause biodiversity loss.

2.1 The Burning of Fossil Fuels and Climate Change

Since the start of the Industrial Revolution in the late 18th century, the extraction and burning of fossil fuels—primarily coal—became the dominant way energy was produced to power everything from steam engines to factories and railways. In the

[4] Daszak et al. (2000), pp. 443–449.

[5] Eguiluz-Gracia et al. (2020), pp. 2170–2184.

[6] Gibbs (2014), pp. 85–91.

mid-19th century, there was a rise in oil and natural gas production to power automobiles, cooking, and heating and cooling buildings. After World War II there was a dramatic increase in the burning of fossil fuels due to a global economic boom. Between 1900–2000 fossil fuel use increased 16-fold, then another 1.5 fold between 2000–2019.[7] Fossil fuel combustion releases greenhouse gases such as carbon dioxide (CO2), methane (CH4), and nitrous oxide (N2O), which accumulate in the atmosphere and contribute to global warming. Concentrations of carbon dioxide, has increased from 365 parts per million (PPM) in 2002 to over 426 PPM in 2024.[8] Methane, another potent GHG, is the second biggest contributor to global warming. It is also emitted through the fossil fuel combustion in addition to agriculture and the decomposition of human and animal waste. Its concentration in the atmosphere has more than doubled in the last 200 years and is responsible for 20–30% of global warming since the Industrial Revolution.[9] The burning of fossil fuels has also led to air pollution, ocean acidification, environmental degradation, a rise in global temperatures, the melting of glaciers, and more intense and more frequent weather events. Oceans, which are a natural carbon sink, have been absorbing excess carbon dioxide, leading to 30% increase in acidity since pre-industrial levels.[10]

From catastrophic hurricanes that cause widespread flooding and property loss, to more frequent droughts that are fueling wildfires, climate change-induced extreme weather events have devastating impacts on the health of people, animals, and the environment. When storms strike, they cause devastating ecosystem damage, water and land contamination, crop failure, and mortality of native species. They increase the risk of aggravated respiratory illnesses from mold, mental health distress, waterborne illnesses, displacement, injury, and death. Hurricanes destroy fragile marine habitats and create new breeding grounds for mosquito-borne illnesses. Extreme drought often results in shortages of water and food, leading to malnutrition and dehydration in humans, plants, and animals. Droughts cause animals to migrate in search of water and food, often throwing off the biological balance of an ecosystem and pushing wildlife into contact with humans, increasing the risk of pathogen spread. Wildfires rip through ancient forests, killing, disrupting, and displacing everything in their path.

2.1.1 Rising Temperatures and Heat Waves

The GHGs trapped in the Earth's atmosphere increase temperatures across the globe. Heat waves have increased in intensity, frequency, and duration. Since the 1950s, cumulative heat has been shown to have increased throughout the world,[11]

[7] Ritchie et al. (2024).

[8] NASA (2024a).

[9] NASA (2024b).

[10] NOAA (2020).

[11] Perkins-Kirkpatrick and Lewis (2020), p. 3357.

and the average number of heat waves per year has gone from two in the 1960s to six in the 2010s and 2020s.[12] Epidemiological studies in humans have shown that heat waves, which can be defined as a period when local excess heat accumulates over a sequence of unusually hot days and nights, are associated with dehydration, heat stroke, aggravated cardiovascular illnesses, aggravated respiratory illnesses, mental health issues, and increased morbidity and mortality.[13] Between 2000 and 2019, it is estimated that 489,000 people were killed each year around the world by extreme heat, making heat the deadliest of all extreme weather events.[14]

Heat affects animals through physiological and behavioral alterations, such as reduced caloric intake, increased water intake, and decreased reproduction and growth. Plants are also negatively impacted by heat through disruptions to photosynthesis, respiration, growth, development, and reproduction.[15] Changes in weather patterns and temperatures can also lead to shifts in migration patterns and breeding seasons. While animals and plants can mitigate some of these effects through adaptive mechanisms such as the generation of heat shock proteins, antioxidants, and others, these mechanisms may be inadequate with further global warming and climate shifts. Prolonged and recurrent heat waves have been shown to lower reproductive rates and increase physiological stresses, leading to species population decline and threatening the future biodiversity of ecosystems.[16]

2.1.2 Air Pollution

Air pollution, which is the presence of harmful chemicals, particulate matter, and pollen in the air, has dramatic impacts on human health. Particulate matter (PM), ground-level ozone (O3), nitrogen dioxide (NO2), sulfur dioxide (SO2), carbon monoxide (CO), and volatile organic compounds (VOCs), are the building blocks of air pollution and are increasing as the result of burning fossil fuels, industrial processes, transportation emissions, forest fires, and other combustibles. The World Health Organization (WHO) estimates that air pollution is responsible for the deaths of 7 million people every year worldwide and 99% of the global population lives in areas in which air quality exceeds the WHO-recommended limit.[17] Longer allergy seasons and increased pollen levels are exacerbating air pollution, leading to more allergies and asthma along with other respiratory illnesses. Wildfires, vehicular and industrial emissions, and an increase in synthetic chemicals are all exacerbating air pollution.

---

[12] US EPA (2024a).

[13] Ebi et al. (2021), pp. 698–708.

[14] Zhao et al. (2021), pp. e415–e425.

[15] Mathur et al. (2014), pp. 116–126.

[16] Mishra (2021), p. 400.

[17] WHO (2022).

Wildfires

Wildfire smoke increases air pollution and has also been shown to have devastating impacts on health. Burning structures release fine particulate matter (2.5 mm and smaller) and other toxic chemicals, which enter the lungs and bloodstream and increase cardiovascular and respiratory diseases, along with other adverse health outcomes.[18] Exposure to toxic air pollution from wildfires is particularly dangerous for people with asthma, kidney disease, heart disease, people who are pregnant, and communities who live closest to the smoke. Additionally, victims of wildfires may experience mental health issues related to being displaced from their homes. Like humans, when animals breathe polluted air they suffer from respiratory illnesses and other reproductive and developmental issues. Acid rain from air pollution damages ecosystems in forests, lakes, and rivers, accelerating habitat loss and altering animals' access to clean and safe food and drinking water. Sulfur and nitrogen from air pollution have shown to have negative impacts on ecosystems through acidification and eutrophication, a process where excessive nutrients, primarily nitrogen and phosphorus, caused rapid algae grown, depleting oxygen in water bodies and degrading both terrestrial and aquatic environments. While the Clean Air Act and subsequent amendments have resulted in steady improvements in ambient air quality in the United States over the past several decades, the increase and severity of recent wildfires have been reversing these trends.[19]

Anthropogenic Pollutants

The rapid increase in anthropogenic pollution—in particular, plastic pollution—is having adverse effects on human, animal, and ecological health. Microplastics, which are <5 mm in diameter, are plastic particles broken down from larger plastic items. They are found in the air, water, and soil, are small enough to penetrate plant tissues, and have been detected in human blood,[20] placenta,[21] stool,[22] and lungs.[23] Plastic pollution, along with heavy metals and other contaminants, are also accumulating in animals, especially those high on the food chain. Fish and other aquatic species absorb contaminants such as microplastics and mercury, which concentrate as they move up the food chain, making them dangerous for consumption by humans and animals. Excess nitrogen and phosphorus can lead to algae blooms, which block sunlight and consume the oxygen in that ecosystem leading to dead zones, where

---

[18] Holder et al. (2020).

[19] Burke et al. (2023), pp. 761–766.

[20] Leslie et al. (2022), p. 107199.

[21] Ragusa et al. (2021), p. 106274.

[22] Schwabl et al. (2019), pp. 453–457.

[23] Pauly et al. (1998), pp. 419–428.

fish and aquatic species can no longer survive.[24] The effects ripple through ecosystems and eventually impact human health. In particular, disadvantaged communities are disproportionately exposed to chronic health issues like asthma, cancer, and cardiovascular diseases due to increased exposure to pollutants.[25] Human culture of overconsumption has led to widespread and catastrophic pollution, further increasing the strain on the natural environment and throwing off the balance of ecosystems, species, and the overall health of the planet.

2.2 *Deforestation, Urbanization, and Loss of Biodiversity*

Deforestation, which is the purposeful clearing of trees in forests, has far-reaching implications for the health and well-being of animals, ecosystems, and humans. When forests are clear-cut to make way for a crop, species experience habitat loss and the soil becomes erosive and depleted of its nutrients. Ramping up around the Industrial Revolution, forests were cleared to make way for mining, agriculture, and urban expansion. Forests are a natural carbon sink as they absorb CO2. Logging not only destroys habitats and displaces species, resulting in decline and extinction of many, it also reduces the ability of the forests to sequester carbon from the atmosphere. With habitat loss and disruptions in weather patterns, animals lack access to water, food, and shelter, forcing them to migrate. As animals and insects migrate, there are more opportunities for pathogens to spread to humans. According to the WHO, 60% of emerging infectious diseases come from animals, making zoonotic disease transmission one of the pillars of One Health. As environments change, so do the way that animals carry and transmit diseases.[26]

Changes in precipitation and temperature have expanded the geographical range and behaviors of many insects, like ticks and mosquitoes. In areas where insects haven't historically been able to survive, they are now thriving, leading to an increase in cases of vector-borne diseases. Vector-borne diseases include malaria, dengue, yellow fever, Zika virus, etc. For example, we are now seeing Zika and dengue fever—spread by mosquitoes—in temperate climates, where they once could only survive in subtropical or tropical climates.[27] The incidence of Lyme disease, the most common vector-borne illness which is spread via ticks, has nearly doubled since the early 1990s and expanded its geographic range.[28] In addition to the new geographic range for animals and insects, there is an increase in the length of months when vector-borne diseases are most likely to be transmitted.

[24] US EPA. The Effects: Dead Zones and Harmful Algal Blooms. Accessed by https://www.epa.gov/nutrientpollution/effects-dead-zones-and-harmful-algal-blooms.

[25] Brulle and Pellow (2006), pp. 103–124.

[26] CDC (2024b).

[27] Semenza and Suk (2018), p. fnx244.

[28] CDC (2024c).

Animals displaced from their natural habitats find refuge in urban areas, creating new disease pathways from animals to humans. Simultaneously, urbanization and the spread of cities fuel the connection between humans and animals. Across the globe, approximately 56% of the population lives in cities, and that number is expected to rise to 70% by 2050 if urbanization trends continue.[29] Climate shocks will give way to more climate migration where the majority of displaced communities end up in cities, further contributing to urbanization.[30] When cities expand and grow, they encroach upon and disrupt natural ecosystems, leading to the transfer of new and re-emerging zoonotic diseases.

2.3 Modern Agricultural Practices

In conjunction with the rise of fossil fuel use, there was a dramatic shift in the 18th and 19th centuries away from traditional subsistence farming—which included intentional crop rotation, manual labor, and natural fertilizers—to modern agricultural practices and the agricultural revolution which transformed how we produce, distribute and consume food. Starting in the early twentieth century, another profound shift in agriculture occurred when new chemical inputs and advanced machinery were introduced, giving rise to industrial agriculture. The Green Revolution, which occurred between the 1940s–1960s, led to advanced irrigation systems and the introduction of high-yield monoculture crops. This was coupled with an increase in fertilizer use, which increased global food production while also increasing the use of monocultures, which has caused extensive environmental damage. Between 1960 to 2000, food grain production doubled from 1 billion tons to 2 billion tons. In contrast, it took almost 10,000 years to reach 1 billion tons.[31]

The world's dependence on modern industrial agricultural practices has created an unsustainable food system with dire conditions for animals, the depletion of soil health, overuse of water, an increase in water and land pollution, loss of biodiversity, overuse of antibiotics, and the release of GHGs that drive climate change. While industrial agriculture has been able to produce more food, it does so at the expense of destroying ecosystems, pushing animals into contact with people, and creating food that is lower in nutrition and quality.[32] Industrial agricultural practices have led to an increase in runoff and an increased need for pesticides and herbicides that have been linked to cancer in humans[33] and the destruction of downstream ecosystems.

[29] World Bank Group (2023).

[30] UNHRC (2018).

[31] Khush (2001), pp. 815–822.

[32] Davis et al. (2004), pp. 669–682.

[33] Cavalier et al. (2023), pp. 879–912.

3 One Health, Health Inequities, and Environmental Justice

Health inequities represent systematic differences in health status among social groups, often stemming from a legacy of historical biases compounded by social-economic factors. These inequities are viewed as unjust and unfair, manifesting in significant disparities such as life expectancy variations based on race and geographic location. Access to medical care—or the lack thereof—has long been a driving force behind the health inequities in our communities. The health impacts of climate change exacerbate these disparities, as individuals with underlying and pre-existing health conditions and limited access to clean environments are disproportionately affected by climate-related events.[34] Extreme weather events caused by climate change impacts everyone, but not everyone experiences these impacts equally. Underserved communities, including people of color, children, pregnant people, the elderly, people living with chronic conditions, and socioeconomically disadvantaged groups have greater vulnerability to the impacts of climate change, which are amplified by social and environmental stressors.

For example, redlining—a discriminatory housing practice instituted by the Home Owners' Loan Corporation following the Great Depression in the United States—contributed to racial segregation and has resulted in neighborhoods experiencing high temperatures due to urban heat islands.[35] Urban heat islands result from a concentration of cars, dark rooftops, and paved streets that trap heat, making cities hotter than surrounding areas. These neighborhoods often have fewer trees and green spaces, leading to hotter temperatures, which disproportionately affect people of color and low-income residents.[36] Studies have shown that formerly redlined neighborhoods are up to 5 degrees Fahrenheit hotter than others, exacerbating health risks during heatwaves, which particularly harm re populations.[37] Additionally, marginalized groups are more likely to live in flood zones and food deserts—areas in which it is difficult to buy affordable and fresh food—further increasing their exposure to climate-related risks and limiting access to nutritious food and essential health services.[38] These vulnerabilities are a result of multiple factors including disparities in healthcare, lower educational status, and systemic racism. Children are particularly vulnerable with 88% of the world's health burden from climate-sensitive outcomes being incurred by children under the age of five.[39] Vulnerable populations such as children, pregnant people, the elderly, those with comorbid conditions, low-income communities, and communities of color are disproportionately impacted by these climate shocks.

---

[34] Kovats and Hajat (2008), pp. 41–55.

[35] Estien et al. (2024), pp. 54–59.

[36] Zhou et al. (2021), pp. 1764–1775.

[37] Hoffman et al. (2020), p. 12.

[38] Chaudhry (2024), p. 100277.

[39] Sheffield and Landrigan (2011), pp. 291–298.

Climate change is expected to intensify existing health inequities. Marginalized communities already burdened by inadequate healthcare and historical inequities are likely to face the most severe health impacts from the climate crisis. Vulnerable populations often face the brunt of environmental degradation caused by deforestation, industrial pollution, and climate change itself. Indigenous communities, for example, frequently find themselves at the forefront of environmental conflicts due to their close relationship and connections to the land and natural resources.[40] This exacerbation of health disparities underscores the urgency of addressing both the root causes of inequities and the adverse effects of environmental degradation.

The Environmental Justice movement emerged in the 1980s in the United States as a response to marginalized communities being disproportionately exposed to environmental hazards such as air pollution, contaminated water, and unsafe living conditions.[41] For example, low income communities and people of color live closer to freeways, exposing them to dangerous levels of air pollution. In Minneapolis, neighborhoods like Phillips, encircled by major highways, suffer from elevated soil lead contamination and respiratory issues caused by vehicle emissions from nearby freeways.[42] These areas also experience increased noise pollution, adding to long-term health burdens such as high blood pressure and cognitive impairments.[43] Globally, similar patterns of environmental injustice can be observed. In India, communities living near coal-fired power plants face severe air and water pollution, leading to health problems such as chronic diseases and waterborne illnesses. In South Africa, residents of informal settlements often face inadequate water management and infrastructure, relying on contaminated communal taps that lead to waterborne diseases such as cholera and diarrhea.[43] This government failure to ensure equitable access to clean water disproportionately affects marginalized communities, worsening health disparities.

The Environmental Justice movement advocates for the fair distribution of environmental benefits and burdens, ensuring that all people, regardless of race, nationality, or income, have access to a healthy environment. It is grounded in principles of equity and inclusion, aiming to rectify systemic inequalities through legal action, policy reforms, and community advocacy to ensure that all communities receive equal protection under environmental laws and have a meaningful voice in decision-making processes.[44] By addressing the unequal burden of environmental degradation on vulnerable populations, the environmental justice movement seeks to create more equitable living conditions and protect public health.

The concept of One Health is integral to understanding the intersections between health inequities and environmental justice. Communities near industrial sites, for example, not only face high exposure to pollutants but also experience increased

---

[40] United Nations (2024).

[41] Bullard (1994).

[42] McDaniel (2020).

[43] Govender et al. (2011), pp. e4–e9.

[44] EPA (2024).

risks of zoonotic diseases, which often emerge in environments where human and animal habitats overlap. The interaction between humans and animals is heightened in these environments due to factors such as waste management practices that attract wildlife and agricultural runoff that creates habitats conducive to disease vectors. Pollutants from industrial activities including emissions from manufacturing plants, chemical spills and improper waste disposal can contaminate air, water, and soil. Industrial operations can disturb ecosystems, leading to altered wildlife behavior and increased opportunities for disease transmissions, as seen in outbreaks of leptospirosis and salmonellosis—diseases linked to contaminated water sources.[45] Additionally, pollutants can compromise immune systems, making both wildlife and human populations more susceptible to infections like avian influenza and hantavirus.[45] This interconnectedness underscores the urgent need for integrated approaches that address both environmental degradation and public health, particularly for marginalized communities that bear the brunt of these compounded risks.[46] An understanding of One Health and the potential risks is only possible with better communication and collaboration between veterinarians, scientists, conservationists, policymakers and the public health sector.

4 Key Policies to Advance Climate Change Solutions

Awareness of climate impacts on people, animals, and ecosystems has been building for the past half-century. Still, comprehensive international and domestic climate policies have only recently been implemented enabling climate solutions to flourish. While the health impacts of climate change are already being experienced around the world, climate policy and environmental regulations to curb impacts are still relatively new.

Table 1 lists the major environmental policies over the past half a century and the impacts on One Health principles and practices. Internationally, the Paris Climate Agreement which was reached in 2015 under the United Nations Framework Convention on Climate Change was a milestone aiming to limit temperature rise to 1.5 °C in 2100 above pre industrial levels. In 2019, the Green Deal was launched as a plan to make Europe the world's first climate-neutral continent by 2050. It also works to ensure that there is at least 55% less net GHG emissions by 2030 compared to 1990 levels and to have an additional 3 billion trees be planted in the EU by 2030. It not only encompasses a wide range of initiatives to reduce greenhouse gas emissions but also promotes sustainable industry and energy and addresses environmental and health issues.

In 2022, the United States passed three monumental pieces of legislation: the Inflation Reduction Act, the CHIPs and Science Act, and the Infrastructure

[45] Esposito et al. (2023).

[46] Rupasinghe et al. (2022), p. 106225.

Table 1 Major environmental policies over the past half a century and the impacts on One Health principles and practices

Key Date	Significant Event	One Health Contribution
1963	Clean Air Act in the United States	This was the first major environmental policy aimed at regulating air pollution nationwide in the United States. By reducing air pollution, it has improved public health, while also benefiting animal and ecosystem health.[a]
1972	The UN Scientific Conference, also known as the First Earth Summit, in Stockholm, Sweden	This conference led to the establishment of the Governing Council of the United Nations Environment Programme (UNEP) aimed at overseeing international environmental activities and supporting sustainable development. It highlighted the connections between environmental degradation and human health and laid the foundation for a holistic approach to addressing global environmental health challenges.[b]
1979	The First World Climate Conference in Geneva, Switzerland	This conference recognized climate change as a serious global issue for the first time and established the World Climate Programme (WCP), emphasizing the need to integrate climate science with health considerations when addressing climate impacts.[c]
1987	Montreal Protocol	This landmark agreement curtailed the use of Chlorofluorocarbons (CFCs), which were depleting the ozone layer and are a potent greenhouse gas (GHG). By protecting the ozone layer and reducing GHGs, it mitigated health risks from increased UV radiation, which humans, animals, and ecosystems benefit from.[d]
1988	Establishment of the Intergovernmental Panel on Climate Change (IPCC)	Formed by the UNEP and World Meteorological Organization (WMO), this group was established to assess current and future climate risks and impacts. The IPCC has since provided critical climate science that informs policies addressing the combined effects of climate change on human, animal, and ecosystem health.[e]
1992	The Earth Summit, Rio de Janeiro, Brazil The Convention on Biological Diversity	At this summit, the United Nations Framework Convention on Climate Change (UNFCCC) was established to create a framework for international climate negotiations. It was the first international treaty aimed at ecosystem conservation and it highlighted the critical role of biodiversity in maintaining ecosystem, human, and animal health.[f]
1997	Kyoto Protocol in Tokyo, Japan	This was the first binding agreement by the UNFCCC that aimed to cut GHG emissions by an average of 5% below 1990 levels by 2012. This protocol addressed climate change's cascading effects on human, animal, and environmental health.[g]

(continued)

Table 1 (continued)

Key Date	Significant Event	One Health Contribution
2004	Global Framework for the Progressive Control of Transboundary Animal Diseases	Organized by the Food and Agriculture Organization (FAO) and the Organization for Animal Health (OIE), this initiative aimed at preventing and controlling transboundary animal diseases (TADs) in humans and livestock. It promoted coordinated efforts to manage animal diseases, thereby reducing health risks and improving ecosystem stability.[h,i]
2007	New Delhi International Ministerial Conference on Avian and Pandemic Influenza	The Food and Agriculture Organization (FAO), World Organisation for Animal Health (WOAH), World Health Organization (WHO), the United Nations Children's Fund (UNICEF), the World Bank, and the UN System Influenza Coordinator (UNSIC) recommended that the international community develop a medium-term strategy to address emerging infectious diseases.[j]
2009	COP15 in Copenhagen, Denmark Copenhagen Accord	This political agreement recognized the need to limit global temperature increase to below 2 °C above pre-industrial levels. It supported financial assistance for climate adaptation and mitigation measures in developing countries that addressed health risks from climate change for both humans and animals.[k]
2010	Convention on Biological Diversity (CBD) and the Nagoya Protocol	This convention aimed to protect biodiversity and promote sustainable resource use and ensure fair access to genetic resources and benefit sharing. It promoted sustainable biodiversity and use of genetic resources which are important in maintaining balanced ecosystems supporting human, animal, and environmental health.[l,m]
2010	COP16 in Cancun, Mexico Cancun Agreement	This conference formalized lingering elements of the Copenhagen Accord and established the Green Climate Fund to support climate mitigation and adaptation in developing countries. It promoted ecosystem-based adaptation strategies and enhanced climate resilience.[n]
2012	One Health Summit in Davos, Switzerland	This was the first One Health summit which resulted in a One Health Action Plan focused on managing health threats, food safety, and security. It aimed to foster collaboration across all sectors of health and innovative solutions to global health challenges.[o]
2015	Paris Climate Accord	The Paris Agreement, organized by the UNFCCC, shifted to limiting the increase in global temperatures to 1.5 °C in 2100 rather than the previously agreed 2 °C. It promoted climate action and reductions to environmental degradation, protecting ecosystems and mitigating climate-related health risks for humans and animals.[p]

(continued)

Table 1 (continued)

Key Date	Significant Event	One Health Contribution
2015	Global Action Plan on Antimicrobial Resistance	This global action plan organized by the World Health Organization (WHO), Food and Agriculture Organization (FAO), and Organization for Animal Health (OIE) aims to curb Antimicrobial Resistance (AMR). It encourages responsible antibiotics use in humans, animals, and agriculture, reducing resistant infections, and protecting the health of humans, animals, and the environment.[q]
2016	Kigali Amendment	This amendment aimed to phase down global hydrofluorocarbons (HFCs) emissions. It helps mitigate climate change by lessening health risks associated with heat and air pollution, benefiting both human and environmental health.[r]
2017	United States One Health Zoonotic Disease Prioritization Process	This policy, led by the Center for Disease Control and Prevention (CDC) aims to prioritize zoonotic diseases that are threats to humans, animals, and the environment. By identifying and prioritizing zoonotic diseases for research and prevention, this policy improves health outcomes across human, animal, and environmental domains.[s]
2019	European Green Deal	This policy framework was introduced to make Europe a carbon-neutral continent by 2050 through clean energy, GHG reduction, and creating a circular economy. It aims to reduce pollution, protect biodiversity and advance sustainable food systems to combat climate change and improve public health.[t]
2021	COP26 in Glasgow, Scotland Glasgow Climate Pact	This led to the creation of the Global Methane Pledge which aims to phase down coal and reduce methane. It reinforced commitments to reduce GHG emissions and enhance climate resilience, thereby addressing interconnected effects of climate change on animal, human, and ecosystem health.[u,v,w]
2022	COP27 in Sharm El Sheikh, Egypt	This conference centered on advancing climate action through climate adaptation and mitigation, while securing financial support for developing countries to combat climate change. It established the Loss and Damage fund to address the climate-induced health impacts in vulnerable nations, fostering collaboration to improve resilience and reduce these risks.[x,y]
2022	United States Inflation Reduction Act (IRA)	This US domestic policy allocated hundreds of billions of dollars for climate mitigation and adaptation solutions, the largest in history. The act promotes environmental sustainability by reducing air pollution and encouraging cleaner energy systems which benefit human, animal, and environmental health.[z]

(continued)

Table 1 (continued)

Key Date	Significant Event	One Health Contribution
2023	COP28 in Dubai, United Arab Emirates	This conference focused on advancing global climate action by addressing adaptation, mitigation, and financial strategies to combat climate change and enhance resilience. It established the Climate and Health Declaration emphasizing the importance of integrating human, animal, and environmental health to reduce emissions and address health inequities.[aa]

[a] US EPA (2024d)
[b] United Nations (1972)
[c] World Meteorological Organization (2020)
[d] US Department of State (1988)
[e] The Intergovernmental Panel on Climate Change (IPCC) (2024)
[f] United Nations (1992)
[g] UN Climate Change (n.d.)
[h] FAO (2024b)
[i] FAO (2024a)
[j] One Health Commission (2007)
[k] Center for Climate and Energy Solutions (2009)
[l] United Nations (2022a)
[m] Convention of Biological Diversity (2015)
[n] Center for Climate and Energy Solutions (2010)
[o] CDC (2024d)
[p] UN Climate Change (2015)
[q] WHO (2015)
[r] US EPA (2024c)
[s] CDC (2024e)
[t] European Green Deal (2024)
[u] UN Climate Change (2021)
[v] United Nations (2021)
[w] IEA (2022)
[x] United Nations (2022b)
[y] UN Climate Press Release (2022)
[z] US Department of Energy (2022)
[aa] UN Climate Change (2023)

Investment and Jobs Act, making the biggest investment in climate change in history. Totaling nearly half a trillion dollars, the three pieces of legislation have been monumental in kick-starting a clean energy economy in the United States. The CHIP's and Science Act aims to fund climate technology research and development to spur clean energy innovation and supports supply chains. The Infrastructure Investment and Jobs Act aims to help technologies scale as they set aside funds for transmission lines and permitting. The Inflation Reduction Act is the "engine" allocating huge funds to cleantech innovations and environmental justice climate mitigation and adaptation solutions.

5 Adaptation and Mitigation

Adverse planetary and health impacts will increase with intensifying climate-related events, thus investments in climate resilience and mitigation strategies are critical to the protection of our communities, animals, and ecosystems now and into the future. While our understanding of how these climate-related disasters will impact humans is robust, we need more collaboration and multisectoral and transdisciplinary partnerships to understand how they will also impact the natural environment, animals, wildlife, livestock, and pets. In addition to mitigation, it is essential that we have adaptation strategies in place. Climate change adaptation refers to the strategic actions taken to help protect the health and well-being of communities and ecosystems in the face of climate change.

5.1 Climate Change Adaptation Solutions

5.1.1 Food Water and Agriculture

Food systems are under threat due to shifting climates, unpredictable extreme weather events, and industrial agricultural practices. Building climate-resilient food systems is an important adaptation measure to ensure that we can meet the demands of our growing global population. This entails planting climate-resilient crop varieties, agroforestry, water-efficient irrigation systems, crop diversification, desalination, and livestock and pasture management. These strategies contribute to a more stable and reliable food system by enhancing resilience to climate variability and reducing environmental impacts. For example, heavy rains in the Northeast of the United States are becoming more frequent, so the use of cover crops is being recommended to prevent soil erosion, limit nutrient runoff, and prevent pests. Simultaneously, farmers in the West are turning to drought-resistant crops like agave to avoid having their harvest at risk of extreme heat and droughts.

5.1.2 Ecosystem Restoration

Protecting and preparing ecosystems for climate change not only benefits the environment but also has significant health advantages for both humans and animals. Wetland and mangrove restoration has been shown to protect against storm surges, flooding, and erosion thus making them more resilient during hurricanes and tropical storms.[47] Building sea walls has been crucial in preventing storm surge flooding in cities like Venice and New York. Coral reef restoration has ramped up in recent years in response to increased coral bleaching from warming oceans, helping to

[47] Asari et al. (2021), pp. 305–322.

revive vibrant aquatic ecosystems. Habitat corridors allow species to resume normal migration patterns with less human interference, while reforestation efforts—including forest management and planting native trees—support fire prevention and management strategies. In Vietnam, for instance, coastal communities faced increasing risks from storm surges and flooding. In response, the government, in collaboration with NGOs, implemented legislation to protect and restore mangrove habitats. This initiative led to reduced coastal erosion, habitat restoration, improved biodiversity, and enhanced livelihoods for local fishermen. These ecosystem-based adaptation strategies underscore the critical intersection between environmental and human health.[48]

5.1.3 Disaster Risk Reduction

As extreme weather events become more prevalent globally, disaster risk reduction (DRR) emerges as a crucial climate adaptation strategy to ensure the safety of people, animals, and the planet. Effective DDR strategies include early warning systems, disaster preparedness plans, flood risk mapping, wildfire plans, heat action plans, disease surveillance, and resilient healthcare systems. Early warning systems offer timely alerts for impending disasters, allowing communities to reduce damage by implementing preventative measures.[49] Disaster preparedness plans enhance community resilience through coordinated emergency responses, flood risk mapping identifies vulnerable areas and informs targeted mitigation efforts, and heat action plans establish protocols to protect populations from increasing heat events. Disease surveillance systems are vital for monitoring vector-borne zoonotic pathogens, allowing for early detection and intervention to prevent diseases from transferring from animals to humans.[50] The COVID-19 pandemic highlighted how a lack of preparedness and early warning systems can be incredibly costly to the health of humans and the environment. Early detection and prevention of vector-borne zoonotic pathogens can stop pathogens from transferring from livestock to humans when properly monitored and an action plan is quickly put into place.

5.1.4 Policy-based Adaptation

Policy-based climate adaptation strategies are essential for protecting community health and well-being in the face of climate change. National and international climate policies play a critical role in enabling communities to adapt to climate impacts. For instance, urban planning and zoning practices should incorporate climate metrics into their considerations to guide strategic decisions about where and

[48] Tinh et al. (2022).

[49] UNDRR (2022).

[50] Sharan et al. (2023), p. 100050.

how to build, minimizing damages from climate shocks. Building smarter, more resilient infrastructure is another key climate adaptation solution, ensuring that structures can withstand severe weather events and minimize damage. Stricter regulations on stormwater systems and smart water management are vital for maintaining access to clean drinking water and protecting aquatic ecosystems from pollution.[51] Institutions such as the Green Climate Fund (GCF) and the Adaptation Fund provide financial resources and support to developing countries in implementing effective climate adaptation solutions. The GCF focuses on low-emission and climate-resilient projects and the Adaptation Fund emphasizes community-based adaptation efforts in vulnerable regions.[52] By integrating policy-based strategies, we can enhance resilience and promote sustainable health outcomes for people, animals, and the environment.

5.2 *Climate Change Mitigation Solutions*

While countries around the world are making progress to prepare themselves for climate change, their pace has not kept up with our rapidly changing climate, thus we must not only take into consideration how to adapt but also how to mitigate the impacts of climate change. Climate mitigation aims to reduce and prevent the overall emissions of greenhouse gasses (GHG) to help slow global warming.

5.2.1 Sustainable Agriculture and Land Use

Sustainable agriculture and land use are key strategies for climate mitigation, promoting ecosystem health and long-term resilience against climate change. Practices such as regenerative agriculture, methane reduction in livestock, reforestation, afforestation, agroforestry, and conservation contribute significantly to environmental sustainability. Regenerative agriculture improves soil health and increases carbon sequestration through methods like cover cropping and no-till farming, while methane reduction in livestock is achieved by improved feeding, manure management, and rotational grazing.[53] Reforestation and afforestation increase carbon sequestration by restoring degraded lands and creating new forests, respectively, while agroforestry integrates trees into agricultural systems, thereby boosting carbon storage. Conservation practices such as minimum soil disturbance and permanent soil cover prevent erosion and promote biodiversity. The 4 per 1000 Initiative, launched at COP21 in Paris in 2015, exemplifies these strategies by aiming to increase global soil organic stocks by 0.4% per year. This initiative supports regen-

[51] Prasad et al. (2022).

[52] Green Climate Fund (2024).

[53] The Nature, C. Food, Climate and Nature Faqs. (2023).

erative agriculture resilience, combats desertification and reduces CO2 emissions, benefiting farmers, the environment, and the global climate.[54] By improving soil health, this initiative contributes to food security, reduces the need for harmful chemicals, and enhances nutrition. It aligns with the One Health approach by integrating sustainable soil management to support ecosystem health and reduce climate-related health risks.

5.2.2 Energy

Energy is a pivotal element in climate change mitigation, crucial for cutting GHG emissions and advancing cleaner alternatives. Transitioning to renewable energy sources such as solar, wind, hydropower, and geothermal energy is essential, as renewables accounted for approximately 29% of global electricity generation in 2023.[55] Enhancing energy efficiency through building upgrades and electrifying the transportation system the adoption of electric vehicles also significantly reduces fossil fuel dependency. Additionally, carbon capture and storage (CCS) technology helps prevent CO2 emissions from reaching the atmosphere by storing it underground.[56] A prominent example of renewable energy implementation is Iceland's use of geothermal energy, which supplies about 85% of the country's electricity and heating needs.[57] This clean energy source not only reduces greenhouse gas emissions and air pollution but also improves air quality and lowers health risks. Iceland's approach illustrates how sustainable energy practices benefit both the environment and public health, demonstrating the interconnectedness central to One Health.

5.2.3 Sustainable Development and Economic Growth

Sustainable and economic growth are increasingly driven by strategies that emphasize environmental responsibility and resource efficiency. The circular economy plays a key role by focusing on minimizing waste and maximizing resource reuse through practices such as recycling and composting, which help divert materials from landfills and return valuable nutrients to the soil.[58] Additionally, waste-to-energy technologies convert waste into usable energy, reducing reliance on fossil fuels and decreasing GHG emissions.[59] Investing in climate-resilient infrastructure

---

[54] The international "4 per 1000" Initiative. Understanding the "4 Per 1000" Initiative in 3'30. Accessed by https://4p1000.org/?lang=en/ (2021).

[55] International Renewable Energy Agency. Statistics Data. Accessed by https://www.irena.org/Data.

[56] International Energy Agency (2023).

[57] Government of Iceland. Energy. Accessed by https://www.government.is/topics/business-and-industry/energy/.

[58] US EPA (2024b).

[59] Alao et al. (2022), p. 100034.

further supports these efforts by ensuring that communities can withstand and adapt to climate impacts, protecting both people and the environment. The creation of greenspaces and the implementation of nature-based solutions contribute to urban resilience by improving air quality, managing stormwater, and providing habitats and food sources for wildlife. A current example is the city of Copenhagen's integration of green infrastructure and waste-to-energy technologies. Copenhagen has one of the world's largest and cleanest waste-to-energy plants that convert about 400,000 tons of municipal waste energy per year.[60] The city's waste-to-energy plants manage waste sustainably and provide clean energy, while its greenspaces improve air quality and public health. This approach not only fosters economic growth through innovation and resource efficiency but also demonstrates the interrelated benefits for human and environmental health, aligning with the One Health concept by addressing the interplay between ecological systems and human well-being.

6 Future Endeavors to Advance One Health

Significant progress has been made in the understanding of the One Health approach, however, there is more "research to policy to practice" work that is needed to prepare and build resilience in a changing climate. These endeavors will require multidisciplinary collaboration, coordination, and communication between health scientists, veterinarians, conservationists, farmers, healthcare practitioners, policymakers, land managers, and international research and education alliances. Box 1 lists recommendations on how to advance One Health to protect people, planet and animals in the face of a changing climate.

Box 1 One Health: Recommendations to advance research, policy, and interdisciplinary practices to improve climate resilience and mitigation in the face of climate change.

- Build multisectoral collaborations between health scientists, veterinarians and conservationists
- Bring together stakeholders from government, academia, industry, and civil society to discuss legal challenges and opportunities for a One Health Approach and develop actionable items
- Research and monitor emerging zoonotic diseases, infectious diseases, and pathogen spillovers
- Research climate impacts on biodiversity
- Research the effects of climate change-associated displacement and migration on human, animal and ecological health
- Advance research on antimicrobial resistance

(*continued*)

[60] Power Technology (2024).

Box 1 (continued)

- Research efficient, low-cost interventions to mitigate the health effects of climate change
- Advance environmental monitoring of habitats, species, and vulnerable ecosystems
- Monitor and standardize outcome measures of impacts of climate change and effects of mitigation and adaptation measures on human, animal, and ecological health
- Translate research findings into policies to protect One Health across the world
- Push forward and implement policy regulations of wildlife trade and wet markets
- Enhance disease surveillance, modeling and early warning systems
- Increase data sharing and integration to advance surveillance and prevention measures
- Expand public health access and infrastructure
- Build climate-resilient healthcare systems
- Educate and train healthcare workers and veterinarians to prepare for climate change-associated healthcare needs in human and animal populations
- Identify vulnerable populations and allocate adequate resources appropriately
- Promote education and engagement on One Health principles
- Scale up sustainable agriculture and food systems
- Promote wildlife conservation and limit habitat destruction

Figure 1 illustrates a Framework for Advancing One Health from Science to Policy. At the core of the One Health framework are the three C's: Coordinating, Collaborating, and Communicating. Encircling this core are the interconnected areas of human health, animal health, and planetary health, representing the holistic approach of One Health. Surrounding this is a broader circle with knowledge inputs to drive science, fueled by the contributions of health scientists, veterinarians, and environmental scientists. The knowledge generated by these experts informs policy, which in turn, influences key implementers, including farmers & livestock workers, land managers, and healthcare practitioners. The bidirectional relationship of both the knowledge inputs and the implementers shows the interdisciplinary work necessary for bringing science to policy in a One Health framework.

7 Conclusion

We are rapidly altering Earth's environment and the human exposome. There is now near international consensus that the adverse effects on planetary, animal, and human health are brought about by human activities, particularly the burning of

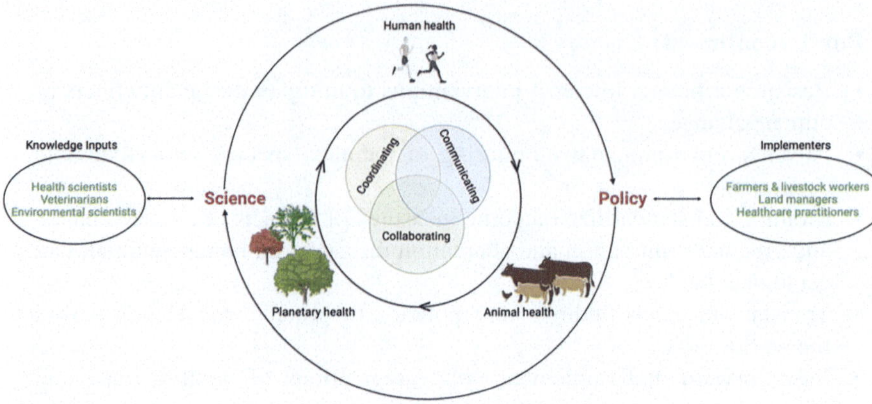

Fig. 1 Framework for Advancing One Health from Science to Policy

fossil fuels. Efforts are being made at an international level to coordinate responses and strategize to combat climate change and increase biodiversity. While talks on the health of the environment have been ongoing since the 1970s, it was not until COP23 in 2023 in Dubai, UAE, that the first Health Day was organized at a UNFCCC. There are critical health challenges posed by zoonoses, antimicrobial resistance, food safety, and climate change, and these require a comprehensive, multi-sectoral approach that breaks down the silos between human health, animal health, and environmental sectors. There is a need to identify gaps and overlaps in the existing laws and regulations that impact One Health activities. The first report in 2023 on the progress of the Paris Agreement meeting its goals warned that governments are not on track to meet the long-term goals of the agreement. International communities need to ramp up efforts to limit further damage to the environment to ensure planetary, animal, and human health are protected.[61] These efforts need to include environmental justice concerns so that adverse human impacts do not disproportionately impact the most vulnerable populations.

References

Alao MA, Popoola OM, Ayodele TR (2022) Waste-to-energy nexus: an overview of technologies and implementation for sustainable development. Cleaner Energy Syst 3:100034

Asari N, Suratman MN, Mohd Ayob NA, Abdul Hamid NH (2021) In: Rastogi RP, Phulwaria M, Gupta DK (eds) Mangroves: ecology, biodiversity and management. Springer, Singapore, pp 305–322

Brulle RJ, Pellow DN (2006) Environmental justice: human health and environmental inequalities. Ann Rev Public Health 27:103–124

Bullard RD (1994) Environmental racism and invisible communities. W Va Law Rev 96

[61] United Nations (2023).

Burke M et al (2023) The contribution of wildfire to Pm2.5 trends in the USA. Nature 622:761–766
Cavalier H, Trasande L, Porta M (2023) Exposures to pesticides and risk of cancer: evaluation of recent epidemiological evidence in humans and paths forward. Int J Cancer 152:879–912
CDC (2024a) About One Health. One Health. https://www.cdc.gov/one-health/about/index.html
CDC (2024b) About Zoonotic Diseases. Accessed by https://www.cdc.gov/one-health/about/about-zoonotic-diseases.html
CDC (2024c) Lyme Disease Surveillance Data. Accessed by www.cdc.gov/lyme/data-research/facts-stats/surveillance-data-1.html
CDC (2024d) One Health History. One Health. Accessed by https://www.cdc.gov/one-health/about/one-health-history.html
CDC (2024e) One Health Zoonotic Disease Prioritization (OHZDP). One Health. Accessed by https://www.cdc.gov/one-health/php/prioritization/index.html
Center for Climate and Energy Solutions (2009) COP 15 Copenhagen. 15th Session of the Conference of the Parties to the United Nations Framework Convention on Climate Change. Accessed by https://www.c2es.org/content/cop-15-copenhagen/
Center for Climate and Energy Solutions (2010) COP 16 Cancun. Center for Climate and Energy Solutions. Accessed by https://www.c2es.org/content/cop-16-cancun/
Chaudhry D (2024) Climate change and health of the urban poor: the role of environmental justice. J Clim Change Health 15:100277
Convention of Biological Diversity (2015) About the Nagoya Protocol. Secretariat of the Convention on Biological Diversity. Accessed by https://www.cbd.int/abs/about
Daszak P, Cunningham AA, Hyatt AD (2000) Emerging infectious diseases of wildlife--threats to biodiversity and human health. Science (New York, NY) 287:443–449
Davis DR, Epp MD, Riordan HD (2004) Changes in usda food composition data for 43 garden crops, 1950 to 1999. J Am Coll Nutr 23:669–682
Ebi KL et al (2021) Hot weather and heat extremes: health risks. The Lancet 398:698–708
Eguiluz-Gracia I et al (2020) The need for clean air: the way air pollution and climate change affect allergic rhinitis and asthma. Allergy 75:2170–2184
EPA (2024), U. Environmental Justice. Accessed by https://www.epa.gov/environmentaljustice
Esposito MM, Turku S, Lehrfield L, Shoman A (2023) The impact of human activities on zoonotic infection transmissions. Animals (Basel) 13
Estien CO, Wilkinson CE, Morello-Frosch R, Schell CJ (2024) Historical redlining is associated with disparities in environmental quality across California. Environ Sci Technol Lett 11:54–59
European Green Deal (2024) Consilium. Accessed by https://www.consilium.europa.eu/en/policies/green-deal/
FAO (2024a) FAO Emergency Centre for Transboundary Animal Diseases (ECTAD). Animal Health. Accessed by https://www.fao.org/animal-health/our-programmes/fao-emergency-centre-for-transboundary-animal-diseases-(ectad)/en
FAO (2024b) The Global Framework for the Progressive Control of Transboundary Animal Diseases (GF-TADs). Accessed by https://www.gf-tads.org/
Gibbs EPJ (2014) The evolution of one health: a decade of progress and challenges for the future. Vet Record 174:85–91
Govender T, Barnes JM, Pieper CH (2011) Contribution of Water Pollution from Inadequate Sanitation and Housing Quality to Diarrheal Disease in Low-Cost Housing Settlements of Cape Town, South Africa. Am J Public Health 101:e4–e9
Government of Iceland. Energy. Accessed by https://www.government.is/topics/business-and-industry/energy/
Green Climate Fund (2024) About Gcf. Accessed by https://www.greenclimate.fund/about
Hoffman JS, Shandas V, Pendleton N (2020) The effects of historical housing policies on resident exposure to intra-urban heat: a study of 108 US Urban areas. Climate 8:12. https://doi.org/10.3390/cli8010012
Holder A, Beidler J, Vukovich V (2020) Estimate of pollutant emissions from fires in the wildland urban interface

IEA (2022) Global Methane Tracker 2022. IEA, Paris https://www.iea.org/reports/global-methane-tracker-2022, Licence: CC BY 4.0
International Energy Agency (2023) Accessed by https://www.iea.org/reports/energy-efficiency-2023
International Renewable Energy Agency. Statistics Data. Accessed by https://www.irena.org/Data
Khush GS (2001) Green revolution: the way forward. Nat Rev Genet 2:815–822
Kovats RS, Hajat S (2008) Heat stress and public health: a critical review. Ann Rev Public Health 29:41–55
Leslie HA et al (2022) Discovery and quantification of plastic particle pollution in human blood. Environ Int 163:107199
Mathur S, Agrawal D, Jajoo A (2014) Photosynthesis: response to high temperature stress. J Photochem Photobiol B Biol 137:116–126
McDaniel T (2020) Environmental Justice. Accessed by https://humantoll35w.org/essays/environmental-justice/
Mishra SR (2021) Behavioural, physiological, neuro-endocrine and molecular responses of cattle against heat stress: an updated review. Trop Anim Health Prod 53:400
Myers SS (2017) Planetary health: protecting human health on a rapidly changing planet. Lancet (London, England) 390:2860–2868
NASA (2024a) Carbon Dioxide. Accessed by https://climate.nasa.gov/vital-signs/carbon-dioxide/?intent=121
NASA (2024b) Methane. Accessed by https://climate.nasa.gov/vial-signs/methane?intent=121
NOAA (2020) Ocean Acidification | National Oceanic and Atmospheric Administration. Accessed by https://www.noaa.gov/education/resource-collections/ocean-coasts/ocean-acidification
One Health Commission. (2007). New Delhi International Ministerial Conference on Avian and Pandemic Influenza. Accessed by https://www.onehealthcommission.org/index.cfm/37526/75803/new_delhi_international_ministerial_conference_on_avian_and_pandemic_influenza
Pauly JL et al (1998) Inhaled cellulosic and plastic fibers found in human lung tissue. Cancer Epidemiol Biomarkers Prev **7**:419–428
Perkins-Kirkpatrick SE, Lewis SC (2020) Increasing trends in regional heatwaves. Nat Commun 11:3357
Power Technology (2024) Amager Bakke Waste-to-Energy Plant. Accessed by https://www.power-technology.com/projects/amager-bakke-waste-energy-plant/
Prasad RLR, Gangireddy M, Mohapatra P (2022) Urban stormwater management for sustainable and resilient measures and practices: a review. Water Sci Technol 85
Ragusa A et al (2021) Plasticenta: first evidence of microplastics in human placenta. Environ Int 146:106274
Ritchie H, Rosado P, Roser M (2024) Fossil fuels. Our World in Data
Rupasinghe R, Chomel BB, Martínez-López B (2022) Climate change and zoonoses: a review of the current status, knowledge gaps, and future trends. Acta Tropica 226:106225
Schwabl P et al (2019) Detection of various microplastics in human stool: a prospective case series. Ann Inter Med 171:453–457
Semenza JC, Suk JE (2018) Vector-borne diseases and climate change: a European perspective. FEMS Microbiol Lett 365:fnx244
Sharan M, Vijay D, Yadav JP, Bedi JS, Dhaka P (2023) Surveillance and response strategies for zoonotic diseases: a comprehensive review. Sci One Health 2:100050
Sheffield PE, Landrigan PJ (2011) Global climate change and children's health: threats and strategies for prevention. Environ Health Perspect 119:291–298
The Intergovernmental Panel on Climate Change (IPCC). (2024) History of the IPCC. Accessed by https://www.ipcc.ch/about/history/
The international "4 per 1000" Initiative. Understanding the "4 Per 1000" Initiative in 3'30 (2021) Accessed by https://4p1000.org/?lang=en/
The Nature (2023) C. Food, Climate and Nature Faqs

Tinh PH et al (2022) Mangrove restoration in vietnamese mekong delta during 2015-2020: achievements and challenges. Front Marine Sci 9

UN Climate Change (2015) Key aspects of the Paris Agreement | UNFCCC. Accessed by https://unfccc.int/most-requested/key-aspects-of-the-paris-agreement

UN Climate Change (2021) The Glasgow Climate Pact – Key Outcomes from COP26. Accessed by https://unfccc.int/process-and-meetings/the-paris-agreement/the-glasgow-climate-pact-key-outcomes-from-cop26

UN Climate Change (2023) COP28 UAE Declaration On Climate And Health. Accessed by https://www.cop28.com/en/cop28-uae-declaration-on-climate-and-health

UN Climate Change (n.d.) What is the Kyoto Protocol? Accessed by https://unfccc.int/kyoto_protocol

UN Climate Press Release (2022) COP27 Reaches Breakthrough Agreement on New "Loss and Damage" Fund for Vulnerable Countries | UNFCCC. Accessed by https://unfccc.int/news/cop27-reaches-breakthrough-agreement-on-new-loss-and-damage-fund-for-vulnerable-countries

UNDRR (2022) Our World at Risk: Transforming Governance for a Resilient Future. Accessed by https://www.undrr.org/gar/gar2022-our-world-risk-gar .

UNHRC (2018) Global Trends Forced Displacement in 2018. Accessed by https://www.unhcr.org/statistics/unhcrstats/5d08d7ee7/unhcr-global-trends-2018.html

United Nations (1972) United Nations Conference on the Human Environment, Stockholm 1972. United Nations; United Nations. Accessed by https://www.un.org/en/conferences/environment/stockholm1972

United Nations (1992) United Nations Conference on Environment and Development, Rio de Janeiro, Brazil, 3-14 June 1992. United Nations. Accessed by https://www.un.org/en/conferences/environment/rio1992

United Nations (2021) COP26: Together for our planet. United Nations. Accessed by https://www.un.org/en/climatechange/cop26

United Nations (2022a) Convention on Biological Diversity, key international instrument for sustainable development. United Nations. Accessed by https://www.un.org/en/observances/biological-diversity-day/convention

United Nations (2022b) COP27: Delivering for people and the planet. United Nations. Accessed by https://www.un.org/en/climatechange/cop27

United Nations (2023) Global Stocktake Reports Highlight Urgent Need for Accelerated Action to Reach Climate Goals. Accessed by https://www.un.org/en/climatechange/global-stocktake-reports-highlight-urgent-need-for-accelerated-action-to-reach-climate-goals

United Nations (2024) The Effects of Climate Change on Indigenous Peoples. Accessed by https://www.un.org/development/desa/indigenouspeoples/climate-change.html

US Department of Energy (2022) Inflation Reduction Act of 2022. Energy.Gov. Accessed by https://www.energy.gov/lpo/inflation-reduction-act-2022

US Department of State (1988) The Montreal Protocol on Substances That Deplete the Ozone Layer—United States Department of State. Accessed by https://www.state.gov/key-topics-office-of-environmental-quality-and-transboundary-issues/the-montreal-protocol-on-substances-that-deplete-the-ozone-layer/

US EPA (2024a) Climate Change Indicators: Heat Waves. Accessed by https://www.epa.gov/climate-indicators/climate-change-indicators-heat-waves

US EPA (2024b) What Is a Circular Economy? Accessed by https://www.epa.gov/circulareconomy/what-circular-economy

US EPA (2024c) Background on HFCs and the AIM Act. Accessed by https://www.epa.gov/climate-hfcs-reduction/background-hfcs-and-aim-act

US EPA (2024d) Summary of the Clean Air Act. Accessed by https://www.epa.gov/laws-regulations/summary-clean-air-act

US EPA. The Effects: Dead Zones and Harmful Algal Blooms. Accessed by https://www.epa.gov/nutrientpollution/effects-dead-zones-and-harmful-algal-blooms

Vora NM et al (2024) Nature-based solutions are essential for climate and health action. The Lancet 0

WHO (2015) Global Action Plan on Antimicrobial Resistan ce. Accessed by http://www.emro.who.int/health-topics/drug-resistance/global-action-plan.html

WHO (2022) Billions of People Still Breathe Unhealthy Air: New Who Data. Accessed by https://www.who.int/news/item/04-04-2022-billions-of-people-still-breathe-unhealthy-air-new-who-data

World Bank Group (2023) Urban Development. Accessed by https://www.worldbank.org/en/topic/urbandevelopment/overview

World Meteorological Organization (2020) World Climate Programme (WCP). Accessed by https://community.wmo.int/en/world-climate-programme-wcp

Zhao Q et al (2021) Global, regional, and national burden of mortality associated with non-optimal ambient temperatures from 2000 to 2019: a three-stage modelling study. Lancet Planet Health 5:e415–e425

Zhou W et al (2021) Urban tree canopy has greater cooling effects in socially vulnerable communities in the us. One Earth 4:1764–1775

Indigenous Knowledges, One Health, and Planetary Health: Building Bridges for the Response to Climate Change

Brittany Bingham, Nicole Redvers, Sean Hillier, and Tamara Riley

Abstract The understanding and appreciation for One Health and Planetary Health movements, research, and practice are increasing alongside a similar increase in climate change-related events. There has additionally been an increase in the dialogue surrounding how Indigenous traditional knowledges (ITK) may connect to or with One Health and Planetary Health responses towards climate change; however, this has often been led and amplified by non-Indigenous peoples and voices. This chapter will therefore review One Health and Planetary Health from an Indigenous perspective with an aim to seek bridging dialogues for more coordinated responses to climate change and other global crises. This chapter will also explore the unique elements of Indigenous Peoples and their knowledges from a strengths-based perspective to situate and platform interconnected narratives in an increasingly siloed world.

B. Bingham
Faculty of Medicine, University of British Columbia, Vancouver, BC, Canada
e-mail: brittany.bingham@ubc.ca

N. Redvers (✉)
Schulich School of Medicine & Dentistry, University of Western Ontario, London, ON, Canada
e-mail: nredvers@uwo.ca

S. Hillier
School of Health Policy & Management, York University, Toronto, ON, Canada
e-mail: shillier@yorku.ca

T. Riley
National Centre for Epidemiology and Population Health, The Australian National University, Canberra, ACT, Australia
e-mail: tamara.riley@anu.edu.au

G. Antonelli et al. (eds.), *Planetary Health - Laws, Policies and Science on the 'One Health' Approach*, https://doi.org/10.1007/978-3-031-90621-3_9

1 Background

The current "code red"[1] for humanity is becoming ever more apparent given the intersecting crises of climate change, biodiversity loss, marine degradation, and pollution. The current code red is additionally platformed on the awareness that six of the nine planetary boundaries have already been transgressed[2]—human beings are at a precipice. The activity of some human communities on the planet have caused devastating harm to the planet through nature disconnected governance, nature disconnected economic systems, nature disconnected education systems—all stemming from nature disconnected value systems. Many human communities have forgotten, and sometimes forcibly suppressed, the understanding that we are completely and utterly dependent on the health of the planet. Many humans have therefore "lost their identity as organisms within a larger system and thus have lost awareness of how to live sustainably with" the planet.[3] The interconnection within Nature of the human species has been stated to be a fundamental determinant of planetary health.[4] With this,

> [e]cological demise points to an impaired human relationship with its inner self (i.e., [as] humans are Nature and not apart from it). In the broader sense, there is evidence of the loss of an ecologically bound cultural identity. The disconnect from Nature manifests as a fragmented and dissociated identity that cannot recognise itself as part of a system, making it easier to project predatory and abusive impulses onto the environment.[5]

With expected further and continuing advancement of the intersecting planetary crises (e.g., climate change), that will continue to have significant impacts on human health, there has been an increase in the presence of fields, concepts, and movements that seek to bridge understandings of health and the environment.[6] Although environmental health-related lineages within academic fields have been more explicitly defined and operationalized in recent decades, they have been rooted within broader understandings of public health that have been in existence for much longer.[7] EcoHealth, One Health, and Planetary Health are examples of more recent concepts that have been realized through varied but interconnected academic traditions, often themselves growing from other academic lineages (see *Box 1*).

---

[1] United Nations (2021).

[2] Richardson et al. (2023), p. eadh2458.

[3] Redvers et al. (2022), pp. e156–e163.

[4] Faerron Guzman et al. (2021), pp. e253–e255.

[5] Redvers et al. (2022), pp. e156–e163.

[6] Charron (2012); Talukder et al. (2024), pp. 66–75.

[7] Ancheta et al. (2021a), pp. 883–885; and Ancheta et al. (2021b), pp. 1131–1133.

Box 1 Environmental health-related field and practice concepts that have evolved within academic settings.

- *EcoHealth*: "Ecohealth is a field of research, education, and practice that adopts systems approaches to promote the health of people, animals, and ecosystems in the context of social and ecological interactions…As a field of scholarship, ecohealth research draws on the natural sciences, health sciences, social sciences, the humanities, and beyond, often working in collaboration with interested parties and community members to address issues at the interface of health, ecosystems, and society".[8]
- *Environmental Health*: "Environmental health centers on the relationships between people and their environment… The environmental health field advances policies and programs that reduce chemical and other environmental exposures in air, water, soil and food".[9]
- *One Health*: One Health is an integrated approach that aims to balance and optimize the health of humans, animals, plants and ecosystems. The One Health approach recognizes that the health of humans, animals, plants and the wider environment are interconnected and interdependent. A One Health approach mobilizes multiple sectors, disciplines and communities to work together and tackle threats to health and ecosystems. More specifically, addressing a collective need for clean water, air, safe and nutritious food and action on climate change and sustainable development.[10]
- *Planetary Health:* Planetary Health is a solutions-oriented, transdisciplinary field and social movement focused on analyzing and addressing the impacts of human disruptions to Earth's natural systems on human health and all life on Earth.[11]

One Health and Planetary Health concepts in particular have been getting increased exposure at regional, national, and international levels. Planetary Health, One Health, and EcoHealth concepts generally focus "on the interconnection between ecosystem protection and human-environment well-being"[12] from varied perspectives. The concepts are somewhat similar in acknowledging the "connections between ecological interactions, community involvement, and the health of people, animals, and the environment".[13] Although rarely acknowledged explicitly within academic publishing directly, in practice, there have been implicit disciplinary

[8] Parkes et al. (2014).

[9] American Public Health Association (2024).

[10] UN Environment Programme (2024).

[11] Planetary Health Alliance (2024).

[12] Talukder et al. (2024), pp. 66–75.

[13] Id.

conflicts between the fields of One Health and Planetary Health. One Health, for example, has been seen to have some structural problems with biomedical reductionistic framing being dominant resulting "in a focus on human health threats from animals" as one main pillar of investigation and practice. Varying additional theoretical frameworks, such as "Relational One Health" have been introduced to try to more explicitly bring in the environmental/ecosystem domain as a key pillar in on-the-ground One Health practice (Meisner et al., 2024). Planetary Health has been criticized on the other hand for being too human orientated with the emphasis "mainly on the consequences for our [human] health through disturbances in the environment".[14] Planetary Health has additionally been framed as a "new" field despite the underlying concepts not being new.[15]

More effort has recently been placed on leveraging the perceived differences between One Health and Planetary Health. Despite both concepts being built on relatively equivalent systemic principles, there continues to be "ongoing debate and emerging confusion around their differences and application areas".[16] Regardless, it has been stated that,

> One Health and Planetary Health are highly complementary fields of scientific inquiry with solid leverage for translation into policy and practice. There is an opportunity to build a stronger research community and transdisciplinary evidence to collectively address pressing public and global health issues in a truly integrated way. Above all, life on Earth is at risk, and our diversity will only make us stronger.[17]

1.1 Indigenous Knowledges

Perceived differences between academic disciplinary concepts (e.g., One Health/Planetary Health) that have been partially or fully conceptualized from within EuroWestern-centric knowledge traditions are arguably of little value to Indigenous Peoples. Indigenous Peoples and their knowledges (see *Box 2*) have from time immemorial been interconnected with plants, animals, ecosystems, and Mother Earth. With fundamental understandings that place humans as just one small part of the larger ecosystem (i.e., cosmocentric or ecocentric worldviews[18]), in comparison to the hierarchical approach of humans and 'certain' animal species being implicitly or explicitly dominant within research focuses (i.e., anthropocentric worldviews), Indigenous Peoples and their Knowledges are well placed to be a grounding point for academic research, policy, and practice.

---

[14] Martens (2024), pp. 128–129 and Martens (2023).

[15] Martens (2023).

[16] de Castañeda et al. (2023), pp. e109–e111.

[17] Id.

[18] Lucero and Cruz (2020), p. 1.

Box 2 Relevant terminologies used in this chapter

- *Indigenous Peoples*: "Indigenous Peoples and Nations are those which, having a historical continuity with pre-invasion and pre-colonial societies that developed on their territories, consider themselves distinct from other sectors of the societies now prevailing on those territories, or parts of them. They form at present non-dominant sectors of society and are determined to preserve, develop and transmit to future generations their ancestral territories, and their ethnic identity, as the basis of their continued existence as peoples, in accordance with their own cultural patterns, social institutions and legal system" (Martínez Cobo, 1981).
- *Indigenous Knowledges/Indigenous Traditional Knowledges*: Indigenous Knowledges are Traditional Knowledges or knowledge that comes from a community. Traditional Knowledges are transmitted from generation to generation; emerge from complete knowledge systems; ere expressed in many formats. eg. oral, ceremony, artistic creations, artifacts, etc.; are not all in the past; there is continued growth, innovation and change in knowledge and practices; includes history, law, spirituality, agriculture, environment, science, medicine, animal behaviour and migration patterns, art, music, dance, craft, construction, and more.[19]
- *Indigenous Knowledge Systems*: Indigenous knowledge systems (IKS) refers to the broader frameworks within which Indigenous knowledges are transmitted and practiced. IKS sets the framing for the cultural, spiritual, and ecological principles that guide how Indigenous knowledge is created, shared, and passed down for future generations. IKS includes the spiritual and cultural frameworks that guide how Indigenous Peoples validate knowledge, ethical frameworks, and collective decision making.
- *Indigenous Planetary Health*: "Planetary health as a 'field' is primarily a Western construct as Indigenous Traditional Knowledge systems have no clear separation between the health of the planet and the health of self or that of the community and the ecosystem at large. This means that the meaning and applications of planetary health are directly rooted in community values based on protocols for living in harmony with all that have existed for thousands of years".[20]

Both One Health and Planetary Health in their conventional framings have been born from EuroWestern-centric academic traditions and worldviews. Their self-embraced interconnectivity between the health of people and the planet (animals, ecosystems, etc. included), have often caused field specific self-declared reflections that they are aligned with Indigenous world views. In other instances there are calls

[19] University of Alberta (2024).

[20] Redvers (2021), p. 678545.

for integrating Traditional Knowledges within the disciplines.[21] These declarations and calls have often not included Indigenous Peoples. Recently, however, there has been increasing interest from Indigenous scholars and communities themselves to examine the foundations of both One Health and Planetary Health from an Indigenous worldview. Therefore, the next sections of this chapter will present an overview of One Health and Planetary Health from 'Indigenous perspectives' with an aim to seek bridging dialogues for more coordinated responses to climate change and other global crises.

2 Positionality

It is increasingly expected within academic publishing in reference to Indigenous Peoples, that authors position themselves in the work.[22] This ensures a more appropriate evaluation of "nothing about Indigenous Peoples, without Indigenous Peoples".[23] With this, our positionality as Indigenous co-authors reflects viewpoints from different contexts and Indigenous territories across Canada and Australia, with our collective work and lived experiences grounding us in our cultures, contexts, and experiential community perspectives. We share commitments towards transformative approaches for Mother Earth's health. BB is a mixed ancestry woman and a member of the shíshálh nation, which is Coast Salish First Nations on the coast of British Columbia, Canada. BB is trained as a PhD public health and community-engaged scholar, researcher, and advocate focusing on Canadian Indigenous Peoples' equity, rights and wellness. BB approaches her work from a decolonial lens and orientation, participating in deep reflection on how Indigenous Knowledges, values, and epistemologies can position us to address the complexities and challenges of planetary health. NR is a member of the Deninu K'ue First Nation located in the sub-Arctic region of Canada. NR is a past-clinician, and Indigenous health scholar with a focus on Indigenous Planetary Health. TR is a Wiradjuri woman with family ties to western New South Wales in Australia. TR is a veterinarian and epidemiologist, and completed her PhD on the implications of One Health for Indigenous communities. TR adopts interdisciplinary One Health research approaches and Indigenous research methodologies to her work that aims to understand current and emerging health challenges for Indigenous populations. SH is of mixed ancestry, a member of the Qalipu First Nation and western European - he was raised on the southwest coast of Newfoundland. SH is a Mi'kmaw scholar with a focus on examining access to wholistic health for Indigenous Peoples and foregrounding Indigenous ways of knowing into One Health research projects.

[21] Padhi and Agarwal (2024).

[22] Lock et al. (2022), p. 7353; Roach and McMillan (2022), p. e0000999.

[23] Id.

3 Indigenous Perspectives of One Health

This section of the chapter will discuss One Health from an Indigenous perspective with discussion of the role of Indigenous Peoples, recognition of Traditional Indigenous Knowledges, and consideration of nature and zoonoses within the global One Health movement.

The One Health High-Level Expert Panel defines One Health as "an integrated, unifying approach that aims to sustainably balance and optimize the health of people, animals, and ecosystems. It recognizes the health of humans, domestic and wild animals, plants, and the wider environment (including ecosystems) are closely linked and interdependent".[24] One Health is promoted internationally as an effective approach to understanding and combatting current and emerging health risks to humans. To date, One Health has been largely focused on how antimicrobial resistance and zoonosis impact human health. A key goal of using a One Health approach is to break down the traditional disciplinary silos in which zoonosis and antimicrobial resistance work has happened, generally within the confines of human health, animal health, or environmental health. However, in line with the definition, it also has a focus outside of human health, encapsulating the wellbeing of all species within ecosystems and the wider environment to address local and global health challenges.

The quadripartite (World Health Organization, Food and Agricultural Organization of the United Nations, World Organisation for Animal Health, the United Nations Environment Programme) supports countries adopting One Health approaches through a One Health joint action plan.[25] The One Health movement, however, has arguably been driven mainly through health organisations based in the Global North. With this, there is a critical need to highlight the relevance of the movement for the many communities outside of these health organizations and networks, including Indigenous communities globally. Only through the dismantling of the disciplinary silos within our research and health systems, alongside the formal challenging of the idea that there is a singular monolithic knowledge system, will we be able to combat inequities in resourcing and the imbalances of power within One Health movements.

In working to advance One Health, there is a need to reexamine how the current definitions of One Health are being taken up by researchers. We, as Indigenous co-authors, based on our experiences, question whether the current understanding of 'One Health' is relatable or applicable outside of people who work within western science. While the overarching goals and approach of the Western-developed definition of One Health could be argued to be aligned with Indigenous worldviews of wholistic health and wellbeing, it does not currently account for the cultural,

[24] Adisasmito et al. (2022), p. e1010537.

[25] Food and Agriculture Organization of the United Nations, United Nations Environment Programme, World Health Organization, World Organisation for Animal Health (2022).

ecological, or spiritual contexts and relationships that are central to Indigenous cultures and communities. In the widespread application of One Health currently, there is a continued foregrounding or prioritization—intentional or not—of human health outcomes. This includes main definitions which state, for example, that One Health seeks to "balance and optimize the health of people [listed as a first priority], animals, and ecosystems".[26] Much of the One Health work completed to date suggests that addressing animal and ecosystem health is being done only in so far as it works to mitigate the risk posed to human health. We continue to see ecosystem health as an area either fully ignored or largely limited within its scope, with a tendency to focus on animal and human interfaces. The implicit foci and end goal of One Health therefore remains the solving and addressing of human health that can be viewed in contrast to Indigenous teachings which promote our connection as stewards and protectors of the environments, ecosystems, waters, and animals. This connection is not done simply to result in our own good health but in recognition of our intimate connection to all things.

A lack of environmental and ecosystem engagement within One Health may be tied to the overall value placed on it by researchers when seeking answers to their questions. In *The Fourth World*, Arthur Manual (2019) offers that the conflict between western society and Indigenous Peoples is a struggle between two competing conceptions of how we view and understand Land. For Western society, land is a commodity to be "speculated, bought, sold, mortgaged, claimed by one state, surrendered or counter claimed by another".[27] For many Indigenous communities the Land is a relationship that gives rise to our practices and ways of life. The notion of Land as a relationship is incompatible with the idea of land as a commodity and creates the struggle between settlers and Indigenous Peoples. It results in a denial of Indigenous Peoples' ability to determine our own futures. Instead, Indigenous communities become what Eileen Moretun Robinson calls a white possession.[28] That Indigenous Peoples are the subjects of white settler majorities within white settler nations, nations that Indigenous Peoples do not claim as our own, and for most of history, refused to claim us. The denial of our conception of Land and the foreclosure of our ability to determine our own future makes us inhabitants of what Manual (2019) called the fourth world.

The implications of Manuals (2019) critique helps us to understand both the limitations and promise of One Health. First, while One Health seeks to include a more wholistic (or de-siloed) understanding of health and its connection to environmental systems, it does not seek to, nor is it capable of, fundamentally transforming the idea of land as a commodity. Without challenging territorial commodification and the ownership and control of Indigenous Lands by non-Indigenous majorities, Indigenous Peoples will always be subject to the externalities of capitalist production which expose us to disease and pollutants. This creates a dualism within One Health that makes it difficult for Indigenous Peoples to fully understand and

[26] Adisasmito et al. (2022), p. e1010537.

[27] Manuel and Posluns (2019), p. 6.

[28] Moreton-Robinson (2015).

embrace. The case of Asubpeeschoseewagong First Nations, commonly referred to as Grassy Narrows, provides an unsettling example of the effects of Land commodification on Indigenous bodies and health. For decades, a pulp mill in the nearby community of Dryden dumped ten tons of mercury-laced effluent into the Wabigoon-English River system.

One Health, as it is currently being used, does not embrace, understand, or account for the inherent spiritual and relational aspects of health and Land. The impacts mercury poison has and will have on humans, animals, and the environment for generations to come for Asubpeeschoseewagong First Nations must also be viewed with a lens of understanding those conditions before, during, and after an incident. As the case above demonstrates, One Health must integrate a temporal lens to the challenges it seeks to solve – by relying on the deep knowledges Indigenous Peoples have of the Lands they steward and how we can regain balance in both the immediate and long term. As an example, the Seventh Generation Principle, based on Haudenosaunee philosophy, states that the decisions we make today should result in a sustainable world seven generations into the future. The Seventh Generation Principle today is generally referred to in regard to decisions being made about our energy, water, and natural resources, and ensuring those decisions are sustainable for seven generations in the future. Maori Peoples of Aotearoa (i.e, New Zealand) have similar concepts to the Seventh Generation Principle. For example, in reciting whakapapa or the genealogy of all things, which is central to all Maori institutions and connects people to the past, present and future and the non-human world—all things are seen to be connected including the air, the water, the land, and the animals. "The Seven Generations Principle cannot be separated from correlating values within Indigenous cultures. These may include stand-alone values such as matrilineal family structures, respect for Elders, relationships with ancestors and preservation of culture; or the principle could be linked to a larger value system".[29] One Health practitioners should aim to address wholistic health challenges with a long-term view into the future.

3.1 *One Health and Zoonosis*

Many Indigenous communities have a strong connection to Land or Country ("the term Land is used in some parts of the world and Country in others"[30]), and have stewarded their Land and Waters for generations through traditional and sustainable practices, including carrying out complex wildlife surveillance for animal management. Despite these long-standing expert knowledges, a continued lack of Indigenous representation and perspectives within One Health persists.

[29] Rousseau-Thomas (2023).

[30] Redvers et al. (2022), pp. e156–e163.

Even though there are increasing numbers of studies utilizing a One Health approach with Indigenous Peoples, there continues to be limited work that has integrated and foregrounded local Indigenous knowledges into the One Health approach.[31] There additionally remains inadequate research that explores the application of One Health to broader Indigenous health and wellbeing outcomes.[32] Despite these issues, there are many opportunities for One Health to learn from Indigenous Traditional Knowledges that have used wholistic and de-siloed interventions for centuries to address threats and challenges to human health, as well as safeguarding and promoting environmental and animal health. Many Indigenous cultures and Traditional Knowledges recognise the deep interconnections, including spiritually, humans have with the Lands, Waters, and Animals we live amongst. As an example, there are many versions of the "Medicine Wheel"(Manitowabi, 2018) teachings passed on within some Indigenous communities in Canada. These teachings vary from one community to another, but the teachings describe the various aspects of life, both seen and unseen. The Medicine Wheel provides teachings about how to be wholistically healthy and the need for us to be in good relations with all living things and the spirit world. The four directions embodied within the Medicine Wheel (Manitowabi, 2018) remind us not only of the need to have wholistic (mental, spiritual, physical and emotional) balance to achieve health in our personal lives but that we must ensure balance (and good health) is maintained within the animals and environments that connect us. We cannot have one without the other. These ideals are transferable to a One Health approach when we view a key function of One Health to be about how to de-silo human, animal, and environmental fields of practice, and the bringing together of people to tackle real world problems with the end goal of restoring true wholistic balance to that problem.

Despite the challenges inherent within One Health, One Health approaches may be particularly helpful for under-resourced communities that live closely with animals and have strong connections to the environment, to combat the health inequities and food insecurity that they often face. Indigenous communities around the world find themselves under-resourced and facing disproportionate health risks across animal, human, and environmental spaces. This includes a lack of access to animal and preventative health care, greater contact with wild animals and vectors, large domestic animal populations, and inadequate/stable housing conditions, among others.[33] Additionally, the impact of the changing climate continues to disproportionately impact Indigenous populations through both exposure risks and widening health inequities.

Awareness of zoonoses has increased over recent years due to multiple outbreaks related to animals, with countries recognising the need to prepare for future

---

[31] Hillier et al. (2021), p. 100248.

[32] Hillier et al. (2021), p. 100248.; Riley et al. (2021), p. 11303.

[33] Ali et al. (2018), p. 2827; Burleigh et al. (2015), pp. 145–150.

outbreaks. Current national health approaches commonly operate within a siloed structure that does not allow for effective communication and collaboration. Importantly, even when comprehensive groups are brought together, Indigenous Peoples are typically forgotten. As the regions around the Arctic Circle continue to warm faster than any other region, Indigenous Peoples are likely to be at the forefront of possible zoonotic infections.[34] Even though Indigenous Peoples continue to steward vast parts of the world, placing them at the forefront of possible risk transmissions, they are generally excluded from surveillance and early knowledge sharing activities. These activities could benefit from Indigenous Knowledges that are deeply rooted in the Land which brings with it a strong understandings of the changes in the environments, animal migration and health, and waters in which communities rely on. In Canada for example, the three northern territories, which are largely rural, represent two-fifths (39.1%) of the country's landmass but 0.3% of its population.[35] Rural and remote populations, largely made up of Indigenous Peoples play a critical role in understanding the past and present conditions, while helping to determine and shape the future when their deep knowledge and usage of the Land is respected and accounted for. De-siloing current health approaches as well as reflecting on how we think about who has and can contribute knowledge to issues or solutions is needed. This requires operating outside of western disciplinary boundaries to truly undertake a One Health approach.

The One Health research approach incorporates components of animal, human, and environmental health within projects; however, when undertaking One Health research with Indigenous populations, the approach should also consider incorporating cultural, ecological, and social factors, as well as utilize respectfully with communities 'Indigenous research methodologies'.[36] Evidence has found the application of One Health with Indigenous communities to be limited and with minimal inclusion of Traditional Knowledge.[37] This gap creates the need for studies that support Indigenous leadership, priorities and viewpoints, to strengthen the evidence base and tailor One Health recommendations to communities.[38] Research that explored zoonoses within Australian Indigenous communities found that there was a small evidence base, that the existing studies mainly investigated the zoonotic pathogens from one health sector, and there was minimal examples of interdisciplinary approaches to zoonotic disease. As zoonoses are a multifaceted health risk, there is a need to consider the contribution of animal, human, and environmental health factors to the impact of zoonoses in Indigenous communities to inform effective disease management.[39]

[34] Hillier et al. (2021), p. 100248.

[35] Statistics Canada (2022).

[36] Rigney (1999), pp. 109–121.

[37] Hillier et al. (2021), p. 100248.

[38] Riley et al. (2021), p. 11303.

[39] Riley et al. (2022), p. e0000921.

Box 3 One Health Case Study
Many Indigenous communities in Australia have large domestic animal populations with companion animals being an important part of community life, Traditional Knowledges, and stories. Yet, many communities face barriers in accessing animal health care leading to intersecting health risks including high rates of disease and welfare concerns. There are examples of addressing these health access barriers by implementing preventative animal health programs. One example presents a community animal health and management program driven by an Indigenous Ranger group in partnership with a not-for-profit organization in remote Australia, which involves assessing animal health and administering preventative health care to dogs and cats within the community.[40] An evaluation of this program found significant benefits to animal health and welfare outcomes with community involvement being an important factor for success. There is now a need to further evaluate if there have been improvements to human and environmental health within similar community programs.

4 Indigenous Perspectives of Planetary Health

This section of the chapter will discuss Planetary Health from an Indigenous perspective with discussion of the interconnectedness of all life, sacred reciprocity and stewardship, decolonizing practices, sovereignty and justice, knowledge co-production, and resilience and adaptation.

As noted previously, intersecting global and ecological crises are increasingly posing a threat to human health and wellbeing. More recent research attention has been focused on the impacts and effects that climate change has on health and wellness, with climate change begin described as one of the greatest public health challenges we will face this century. Aside from the *Box 1* definition (see *Box 1*), Planetary Health as a concept has been defined and understood as "the achievement of the highest attainable standard of health, wellbeing, and equity world-wide through judicious attention to human systems – political, economic, and social – that shape the future of humanity and the Earth's natural systems that define the safe environmental limits within which humanity can flourish".[41] There has been criticism, however, that the current field of Planetary Health has predominantly relied on Euro-Western-centric ideas of health and wellbeing that are inherently separated from natural systems. Over recent years, Indigenous scholarship has drawn focus to how important and essential Indigenous Knowledges are in the discussions surrounding Planetary Health. Indigenous understandings of the universe, including the Traditional Knowledges passed down through generations since time immemorial are pivotal to ensuring we have a healthy planet in the future.

[40] Riley et al. (2020), p. 1790.

[41] Whitmee et al. (2015), pp. 1973–2028.

The Lancet Commission on Planetary Health in 2015, concluded that the,

> ...present systems of governance and organization of human knowledge are inadequate to address the threats to planetary health. We call for improved governance to aid in the integration of social, economic and environmental policies and for the creation, synthesis, and application of interdisciplinary knowledge to strengthen planetary health (Whitmee et al. 2015).

Additionally, scholars have increasingly called for greater consideration of the power dynamics including the critical analysis of climate change frameworks to consider colonization and contemporary processes of environmental dispossession, exploitation and ecocide, as well as considering climate change as a process that has been imposed on Indigenous Peoples through colonialism.[42] Redvers and colleagues in 2022, brought together a group of Indigenous scholars, practitioners, Land and Water defenders, Elders and Knowledge Holders to envision and define "the determinants of planetary health" from an Indigenous perspective.[43] In consultation with Indigenous leaders globally, the determinants of planetary health were described under three broader categories: Mother Earth level determinants, interconnecting determinants, and Indigenous People's level determinants (see Fig. 1). Mother Earth level determinants encompass respect for the feminine and the notion that women are the stewards of the natural environment and cultural identities; with ancestral personhood designation referring to the protection of both human and non-human relatives.[44] Interconnecting determinants refers to the human interconnectedness with nature; and self, and community relationships. Other interconnecting determinants include the modern scientific paradigm, referring to the dominance of western scientific methods, and the need to appreciate Indigenous science and methods for understanding the complexity of planetary health and the world.

4.1 *Interconnectedness of Life*

The scholarly literature is beginning to catch up to the knowledges that Indigenous Peoples have held for millennia about the intrinsic interconnectedness of all living things on Mother Earth. Yet this understanding and appreciation for the

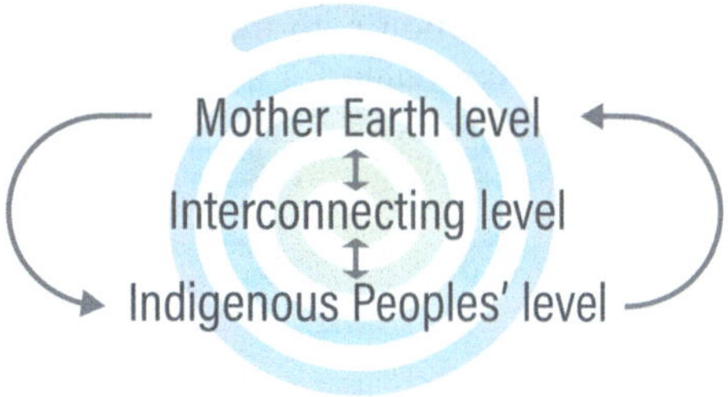

Fig. 1 The interconnectedness of the determinants of planetary health (*used with permission from Redvers et al.* 2022)

[42] Whyte (2017), pp. 153–162.

[43] Redvers et al. (2022), pp. e156–e163.

[44] Id.

interconnectedness of all living things is not often explicitly discussed but is just known inherently within many Indigenous communities. Indigenous Knowledges are Land-based and rooted within spiritual values, as opposed to any kind of materialistic value as seen within a capitalist system.[45] Indigenous Knowledges and ways of seeing the world offer a deeply important way to view Planetary Health and accompanying climate change action.

Planetary Health frameworks have been predominantly anthropocentric (i.e., rooted in how earth's systems can benefit human needs), whereas Indigenous frameworks and concepts are rooted within the intrinsic value of Mother Earth and her inherent worth that exists regardless of human habitation. Natural Law or First Law refers, in the English language, to the guiding principles of many Indigenous Peoples in how our planet is governed through maintaining respectful and reciprocal relations with our non-human relatives and all living things on Mother Earth.[46] This deep personal relationship between all human and more than human beings, and the consequent dynamic balance of life is the essence of the Law. This Natural or First Law, as taught through generations of stories and passed down through generations, creates an ethical framework to live by, for maintaining peace, for thriving, and for living within spaces of love and reciprocity.[47] Indigenous Peoples are from the oldest societies in the world and ancient wisdom has been maintained through storytelling across generations. Stories of the interconnectedness of our world are in stark contrast to colonial laws. Colonial laws privilege human communities over the needs of Mother Earth and all other relatives, which serves to diminish Indigenous Natural or First Law that platforms innately a sense of responsibility, reciprocity, and sacred authority.

Natural or First Law are also Indigenous rules or protocols that are grounded in ceremonial action. Indigenous Peoples follow these protocols to ensure an ongoing and sustainable relationship with Mother Earth. The "Seven Grandfather Teachings"[48] are an example of principles and concepts to live a good life with concepts like respect, love, truth, bravery, wisdom, generosity and humility embodied within the teachings passed down. Within many Indigenous perspectives, Land-based does not mean that the Land is separate from or disconnected from ourselves as people. The Land is active in our healing processes, and it is well known within Indigenous communities, that the health benefits that can arise from our relationship and connection with the land are profound.[49] The value of humility becomes increasingly important for us as humans to acknowledge that we are just one small part of the greater system and web of Mother Earth.

---

[45] Redvers (2018), p. 30.

[46] Redvers et al. (2020), pp. 90–107.

[47] Id.

[48] Seven Generations Education Institute (2021).

[49] Redvers (2020), pp. 90–107.

4.2 *Decolonizing Practices, Sacred Responsibility, and Stewardship*

> Elders are the most brilliant scientists of them all. Their understanding of the universe as described in their powerful and symbolic stories provides clues to their deep knowledge. It is our job to listen to those stories and interpret their meanings so that we do not lose the lessons of the past and can ensure better hope for a healthy planetary future.[50]

Indigenous Peoples, their voices, and knowledges have historically been overlooked in climate dialogue and international action. There is an increasing recognition of the value that Indigenous Knowledges bring to climate mitigation, adaptation, and action. Indigenous ways of knowing are expressed through how Indigenous Peoples see the world and the interconnection with the Land is expressed through language, song, stories, traditional medicines, and songs. Through traditional storytelling, the youngest generations learn from Elders how to interact with the environment and how to respect Mother Earth. Values of reciprocity embedded within the relationships with the Land are deeply intertwined in cultural practices and ceremony. When you take something from the Land, such as a medicine, it is protocol to ask permission for this gift. Some Indigenous groups will therefore gift back to Mother Earth through leaving something in return, as this is a common understanding of achieving balance in nature by acting in reciprocity.[51]

Climate change has been platformed as a form of colonialism that is imposed on Indigenous Peoples globally. Colonialism is about domination in many different forms, with the often exploitation of and extraction from Indigenous lands paving the way for capitalist systems to thrive.[52] Whyte (2017), highlights three key themes to Indigenous inquiries into climate change: (1) The anthropogenic (human-caused) climate change as imposed on Indigenous Peoples by historical and ongoing colonization; (2) Renewing Indigenous Knowledges, specifically Traditional Ecological Knowledges (TEK), and strengthening Indigenous self-determination in climate change action; and, (3) Indigenous visions of climate futures as deeply connecting to ancestral histories to adapt to environmental change and reckoning with the disruptions of ongoing practices due to colonialism and capitalism. Some scholars have discussed climate change as a new era of colonization with the risk of climate refugees having to flee climate events. With this, we are now in a stage of colonially-driven forced adaptation to the ongoing changes of environments.[53]

The term Traditional Ecological Knowledges (TEK) refers to the transmission of Indigenous traditional knowledge of Land stewardship principles that are transmitted through oral traditions between generations and interconnected with all other aspects of Indigenous ways of being. TEK is not homogenous and is diverse across

---

[50] Redvers (2018), p. 30.

[51] Yellow Cloud and Redvers (2023), p. 11786302231157507.

[52] Whyte (2017), pp. 153–162.

[53] Id.

the globe, with common elements of being grounded within the Lands on which people are ancestrally tied to. TEK comes in many forms is inclusive for Indigenous Peoples with their relationships to animals and the environment. See *Box 4* below for an example of Indigenous relationships and understandings of TEK through daily lived experience for Iñupiaq Peoples.

The "Indigenous Determinants of Planetary Health" additionally highlights the importance of the "respect for the feminine" in our considerations of Mother Earth but also the importance and the acknowledgment that women in communities are

Box 4 Planetary Health Case Study

Sakakibira (2009), discusses her work with the Iñupiaq Peoples of the Arctic and the centrality of whales to cultural identity, as well as the interconnections of climate change to the spiritual and cultural practices of whaling. The Iñupiaq live an active subsistence lifestyle with year-round activities such as fishing, caribou hunting, sealing and whaling in the spring. There is meticulous preparation for subsistence, and the oneness between humans and the whales is acknowledged as the centre of life, grounded in ritual, ceremony, song, and drumming.[54] Sakakibira (2009) asserts that as climate change leads to changes to the natural environments of the Iñupiaq Peoples, performances and songs are used to express concerns about and adapt to these environmental changes. The Iñupiaq Peoples are so closely spiritually and culturally connected to whales, and this intimate connection is central to their cultural ceremonies, songs, drumming, and language. Sakakabira refers to this intimate connection as whale-consciousness or "cetaceousnes".[55]

Changes to climate and environmental degradation in this Arctic region threaten not only the whale hunt and subsistence, but the survival of these cultural protocols and the interconnectness to all ways of life for the Iñupiaq. For example, drum ceremonies have been central to whaling season, and these cultural protocols are now also used to express challenges, adaptations, and survival despite environmental change. The ceremonies invite back reciprocal relations between humans and the lived environment. "To the Iñupiaq, the drumbeat is the heartbeat of the people of the whales, and the instrument now has the power to reestablish the communication between the Iñupiaq and the whales".[56] The story of the Iñupiaq and the intimate connection with whales and cultural protocols illustrates sacred stewardship responsibility for the planet and the deep interconnections Indigenous knowledge has with the planet.

---

[54] Sakakibara (2009), pp. 289–303.

[55] Id.

[56] Id.

often the caretakers of culture and stewards of Land.[57] Scholars also highlight that climate change has a gendered impact, affecting Indigenous women more acutely as with all other colonial policies. Industries operating on Indigenous Lands pose greater security threats to vulnerable groups of Indigenous women, such as from oil production fields with threats of human trafficking in highly concentrated areas of industrial activity.[58] The importance of gender in systems of dominance and oppression must also be acknowledged in the field of Planetary Health. Oppression of Indigenous women was an instrument of colonization over hundreds of years and now as Indigenous territories deal with environmental change the leadership of Indigenous women must be elevated, acknowledged, and respected.

Hill et al. (2024) reflect on how rematriation is a key component that must be considered in the climate and environmental justice movements. Rematriation could be described as work to restore the sacred relationships between Indigenous Peoples and their ancestral land, honoring matrilineal societies, and being in opposition to patriarchal violence and dynamics (Sogorea Te' Land Trust, n.d.). Rematriation is central to the field of climate justice and to dismantling colonial systems which diminish matriarch roles in communities as knowledge holders and Land stewards who transmit cultural protocols to next generations. Indigenous Peoples Land-based traditions, such as the placenta burial ceremonies as an example, is an active practice grounded in kinship, and is led by matriarchs as a symbol of sacred connection to the spiritual and natural worlds.[59] Reclaiming ancestral Land-based practices led by matriarchs is key to healing and is vital to the sovereignty of Indigenous communities as matrilineal societies. Hill et al. (2024) explore the process of birthing justice and rematriation in the context of climate justice as an act of returning to the sacred mother. Rematriation is a transformative path towards Indigenous sovereignty and privileging Indigenous epistemologies and ontologies, honouring the sacred knowledge that is required to protect the water and the land, and to maintain access to sacred ceremonial practices.

> For Indigenous Peoples, the land is an ancestor who carries our memories, our stories, our traumas. Healing the land is inextricably linked with healing our people.[60]

4.3 Indigenous Self-determination, Sovereignty, and Nature Justice

Hundreds of years of colonization has disrupted Indigenous Peoples' relationship with their ancestral territories and as a result, has in some cased fractured the transmission of Indigenous Knowledges. TEK is the embodiment of Indigenous Peoples'

[57] Redvers et al. (2022), pp. e156–e163.

[58] Whyte (2017), pp. 153–162.

[59] Hill et al. (2024), p. 100314.

[60] Hill et al. (2024), p. 100314.

relationship to the land so when TEK transmission is prevented, it results in potentially great loss for communities.[61] Restoring ecosystems that have been damaged is an enormous task globally, but global governing bodies are beginning to acknowledge the necessity of Indigenous voices being at the table, and seeing the value of Indigenous Knowledges in managing natural resources.

As previously described, the ancestral personhood designation is a key determinant of Indigenous planetary health. A key example of this is the Whanganui River Claims Settlement, which granted the river legal personhood.[62] This was a key shift from identifying a river as an environmental object to instead identifying the river as a legal subject in and of itself.[63] "This argument – known as "nature jurisprudence" – posits that with legal personhood of the environment comes all the protection due under national and international statute".[64] Further in 2017, the Yarra River was legally recognized as a living entity in Australia.[65] These rivers are being acknowledged as imperative to Indigenous culture and inspires ideas for uplifting ancestral personhood and protecting our ancestral beings.[66] Nature jurisprudence has been seen throughout Indigenous Knowledge Systems where Indigenous Peoples refer to governance through their ancestral connections, and that this nature responsibility is a ultimately an overall personal and inherent responsibility to Mother Earth. The rights of nature for all beings on the planet is clearly understood and appreciated within TEK where relational approaches are applied as opposed to the transactional models currently seen in capitalist and extractive frameworks.[67] These nature rights legal cases have significant importance for future protection of animal relatives. More recently a movement to protect the Bison in North America has gained momentum with nations coming together to develop legal instruments to protect the Bison from a rights-based approach, acknowledging their value not only to the Land but as a cultural relative for generations.[68] These examples highlight the importance of Indigenous self-determination and privileging TEK as being key in protecting the Land and all of the more than human relatives.

---

[61] Robinson et al. (2021), p. e13381.

[62] Waitangi Tribunal (2019).

[63] Id.

[64] McDonough (2019), p. 143.; Id.

[65] Yarra River Protection Ministerial Advisory Committee (2016).

[66] Redvers et al. (2022), pp. e156–e163.

[67] Robinson et al. (2021), p. e13381.

[68] Id.

4.4 *Knowledge Co-Production: Education, Research, and the Shifting of Discourses*

The value and importance of embedding Indigenous Knowledges and values into Planetary Health initiatives are clear. There is a need to envision new ways to educate others to begin to move beyond status quo thinking and purely biomedical approaches to learning and training that can envision innovative solutions to the challenges that continue to put Mother Earth at risk. It is increasingly imperative to have Indigenous leadership in education spaces to bring learners and educators on a transformative journey to integrate Indigenous approaches into Planetary Health education. In 2021, the planetary health education framework was developed based on a task force which sought to build on 12 cross cutting principles for Planetary Health education, and bridge language across all disciplines and sectors.[69] The framework considers five intertwined domains at the core of the framework including: interconnection within nature, the Anthropocene and health, systems thinking and complexity, equity and justice, and movement building and systems change.[70] This framework serves to shift systems away from siloed approaches to education and serves as a guiding framework for transformative approaches to education in Planetary Health (see Fig. 2).

Additionally, the importance of collaboration between western and Indigenous Healing and Knowledge Systems is more amplified with the climate challenges ahead. Decolonizing research and methodologies are necessary to ensure that Indigenous voices are listened to and heard in Planetary Health policy, research, and practice. Health professionals and researchers are often not prepared or adequately trained to conduct work that responds to climate impacts, including being unprepared for the shift and transition to environmentally sustainable health care systems. There is a need for universities and educational institutions to embed Planetary Health training across their programming, particularly in the health professions. Education and research movements, however, need to be inclusive of educational approaches that develop students' knowledge and critical appraisal of decolonial world views as it pertains to our Mother Earth. Innovative educational approaches are therefore needed to encourage learners to better understand varied ways of understanding Planetary Health inclusive of embedded Indigenous Knowledges and approaches.[71]

Indigenous languages themselves are deeply rooted and interconnected to the planet and are like a blueprint of Traditional Ecological Knowledges for Indigenous Peoples.[72] Indigenous languages embody teachings about the Land inclusive of understandings of Natural Laws that often do not even have a translation to the English language. These language teachings are held in the hearts and minds of Elders to be transmitted to the next generations. Therefore, protecting the planet and

[69] Faerron Guzman et al. (2021), pp. e253–e255.

[70] Id.

[71] Brand et al. (2023), pp. e97–e102.

[72] Redvers et al. (2023a), pp. e4–e5.

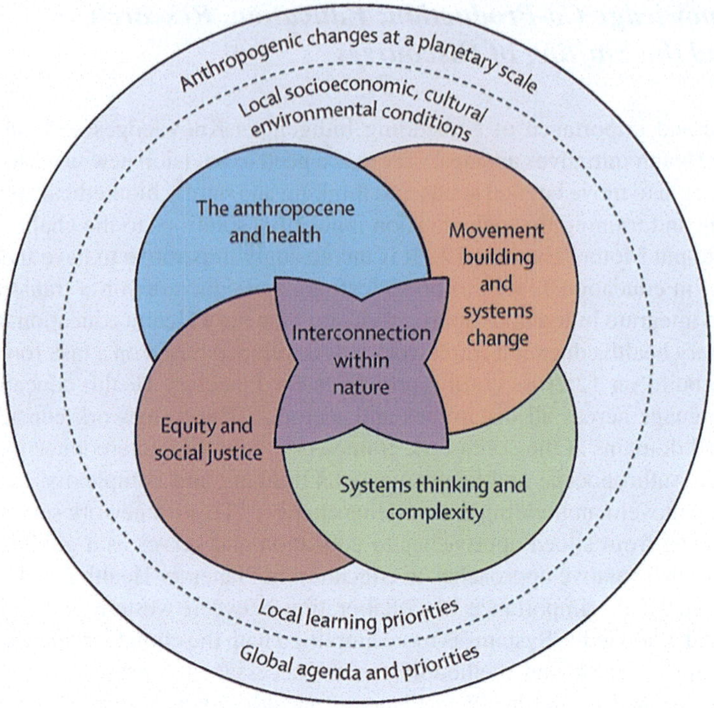

Fig. 2 The planetary health education framework (from Faerron Guzman et al. 2021)

privileging Indigenous Knowledges in Planetary Health also requires the protection and privileging of Indigenous languages. It has been documented that Indigenous Peoples speak more than 4000 of the world's languages but make up less than 6% of the global population.[73] As languages and Traditional Ecological Knowledges are so deeply interconnected, the loss of Indigenous languages due to colonialism has a direct impact on the protection of biodiversity on the planet. Redvers et al. (2023b), highlight that the protection of Indigenous languages is a fundamental determinant of planetary health.[74] Sarah Child additionally noted in 2022 that, "Indigenous languages are the medicine for our people and their revitalization is vital to our individual and collective wellness and the wellness of the earth" (Black Press Media Staff, 2022).

[73] The United Nations Permanent Forum on Indigenous Issues (n.d.).

[74] Redvers et al. (2023b), p. e0002474.

4.5 *Resilience and Adaptation in the Face of Climate Change*

As mentioned previously, Land dispossession as an act of colonization has threatened Indigenous Land-based knowledges, language, kinship systems, and relationality to ancestral lands. Indigenous scholars and climate activists have increasingly recognized the role of Indigenous communities in the protection of Lands and ecosystem stewardship in the face of ongoing colonialism and dispossession. It is often cited that Indigenous Peoples steward eighty percent of global biodiversity while only inhabiting twenty-two percent of the earth's surface and comprising under six percent of the global population.[75] Such numbers are powerful illustrations that support the recognition of the importance of Indigenous Peoples' rights to their lands, as well as the need for continued support for self-determination so they can continue to act as stewards of the living systems on their ancestral territories. Indigenous stewardship of Mother Earth ultimately benefits everyone.

In the face of climate change-related wildfires that are increasingly devastating communities in North America and across the globe, there is increasing focus on the reclamation of Indigenous Traditional Knowledges in fire management. Indigenous Peoples located in North American have Traditional Knowledge not only about fire prevention but also about how fires effect ecosystems, habitats, and resources. Indigenous fire stewardship practices have been maintained for millennia and there continues to be keepers of fire knowledge within many Indigenous communities.[76] Indigenous communities in Canada have had high intensity wildfires over recent years that have often been located in areas that directly impact Indigenous Peoples and threaten Indigenous territories and cultures. Hoffman et al. (2022), states that over 60% of Indigenous communities in Canada are in remote and forested areas and more than 30% are likely to be directly impacted or displaced by the outcomes of wildfires.[77] Cultural fire regimes were traditionally used to affect when, where, and how fires burned through Indigenous territories drawing on passed down cultural knowledge.[78] In Canada, First Nations Peoples regained the right to cultural burning on their lands and territories (it was banned by colonial governments for several decades) however, there remains significant tensions with the need for effective relationships and co-decision making with government oversight and wildfire management agencies.

---

[75] Tauli-Corpuz (2016).

[76] Hoffman et al. (2022), pp. 464–481.

[77] Id.

[78] Lake et al. (2017), pp. 343–353.

Box 5 Case Study: Regenerating and Reviving Indigenous Knowledge in Fire Management

The Tsilhqot'in Peoples of British Columbia, Canada, have been disproportionately and devastatingly impacted by wildfires in their territories (particularly the wildfires of 2017). When the 2017 wildfires broke out in Western Canada, in the interior of British Columbia, over 1.2 million hectares of land was burnt which forced more than 65,000 people to evacuate. The Tsilhqot'in communities were impacted with two of the largest wildfires in the region affecting about 761,000 hectares of land. It then became a key priority for the communities to recover and prevent further damage. These fires occurred only several years after the Tsilhqot'in's Indigenous title was recognized in the Supreme Court of Canada. Despite this court decision, and assertion of jurisdiction over Indigenous territory, there remain challenges in reconciling land authority with the government of British Columbia. Emergency management, for example, remains an area that is managed by multiple jurisdictions from local, regional, provincial, and national bodies. Complex relationships between communities and governments creates jurisdictional complexity in the face of wildfire. Chief Joe Alphonse stated that, "[o]ur traditional, ecological and territorial knowledge of the Tsilhqot'in territory weaves an unbreakable thread through our generations for century upon century".[79]

In 2019, a pilot project was initiated to develop a traditional fire management program for the Tsilhqot'in lands led by the Yunesit'in and Xeni Gwet'in First Nations—both communities being within the Tsilhqot'in Nation. "[T]he Tsilhqot'in word for fire translates to "lightening the load off the land" an Elder shared".[80] Indigenous Peoples have a long history of using fire for their cultural practices and to manage fire risk on their territories. It is something, as noted previously, that has been done for generations and was suppressed as a practice through colonization. The Tsilhqot'in aim to revive and reinvigorate these practices of cultural burning as an integral path forward to prevent the devastation that happened in 2017. Cultural burns are different from controlled burns which are often done by government forest management services. These government led burns are done to address the built-up of fuel within the forests. Cultural burns on the other hand aim to promote growth of medicines, food plants, and fortify lands for all species to thrive. With the Indigenous led project, the Tsilhqot'in Peoples are leading the way in the revival and resurgence of traditional fire management approaches to address the urgent need within their communities. This project is a key example of drawing on the strengths of Indigenous Knowledge and community action for planetary health.

---

[79] Stacey et al. (2019).

[80] Boutsalis (2020).

> Historical conceptions of the boreal forest as a wilderness empty of complex human societies misses both the role that Indigenous peoples played in the boreal landscape, and in particular, the human hand which influences the expression of fire on the landscape and human relationships with other nonhuman things.[81]

Christianson et al. (2022) highlight the deep place-based Indigenous Knowledges of fire and wildland fire management, and that there are currently missed opportunities to bridge this knowledge into current land stewardship and wildfire management practices.[82] Conceptions of the forests being 'empty' landscapes that were untended was commonly used as a settler-colonial justification for claiming empty land through the "doctrine of discovery" when it is now acknowledged that Indigenous Peoples have intentionally tended to the land for generations, often through fire practices.[83] This viewpoint led to the further dispossession of Indigenous Peoples from their Land, disconnecting them from their inherent stewardship duties which were conducted for thousands of years.

Christianson et al. (2022) further explore the Indigenous beliefs that fire is a life-giving force and that everything is interconnected inclusive of fire. Fire is considered sacred to many Indigenous ceremonies and aspects of the culture, however, fire can also be seen as a force that is capable of growth, destruction, and then renewal. Christianson describes how many different Indigenous groups used techniques of fire and burning based on their Traditional Ecological Knowledge. For example, in the Labrador region of Canada, burns were conducted to improve hunting conditions, to create habitat for biodiversity, and to change migration patterns for caribou herd movements. In Northeastern British Columbia, fire is used for clearing vegetation, spiritual and ceremonial hunting, protection from animals and insects, and to improve bison habitat in the territory.[84] It is apparent that cultural burning practices and fire management-related Traditional Ecological Knowledges have been in existence for generations and fractured by ongoing colonialism. Efforts to revitalize fire practices are closely tied to the United Nations Declaration on the Rights of Indigenous Peoples (UNDRIP).[85] With climate change, the impact on forests and increased risks to Indigenous communities' health results in a call for increased Indigenous leadership in fire management and contemporary decision making on Indigenous lands. Finally, Indigenous leadership in fire management systems are increasingly essential to ensure that Land-based traditional and cultural sites of significance are protected for future generations.

---

[81] Christianson et al. (2022), pp. 257–276.

[82] Id.

[83] Id.

[84] Id.

[85] United Nations (2007).

5 Bridging Dialogues to Pave the Way Forward

The tensions and complexities of bridging ways of knowing (e.g., Indigenous and Western) is apparent alongside the urgent need of addressing the health of the planet from climate-related impacts. We must move beyond short-term visions and challenge the dominant knowledge systems and ideological frameworks while acknowledging the importance of Indigenous wisdoms and relational practices. Health systems cannot function independently from the understanding that without the health of people and the planet, we will not have functioning health systems or be able to adequately serve communities. Furthermore, the structural fault lines of inequities will widen as we begin to experience more and more climate events. We must re-examine the values that are held within burgeoning movements such as One Health and Planetary Health and consider and reflect deeply on how we value the planet. There is a definite need for more closer alignment with Indigenous values of relationality and planetary healing.[86] Redvers (2024) poses questions for re-examining core values of healthcare systems: "How do we value the planet? How do our own value systems impact the planet? How do we align our values with relational care of people and planet?".[87] One Health and Planetary Health need to establish a firm interconnection to the Land and Country, and to Land and Country-based principles, resisting the siloed nature of academic fields connecting to the micro, meso, and macro realities that impact the planet. As Indigenous scholars, we highlight in *Box 6* some key recommendation below for moving One Health and Planetary Health forward towards more respectful engagement and inclusion of Indigenous voices.

Box 6 Key recommendations to move One Health and Planetary Health forward towards more respectful engagement and inclusion of Indigenous voices.

- Acknowledge the historical and ongoing epistemicide (i.e., killing of knowledges) of Indigenous Traditional Knowledges within EuroWestern science traditions,[88] that has been present in both One Health and Planetary Health fields.
- Work to uplift Indigenous Peoples, their voices, and their knowledges within One Health and Planetary Health spaces.
- Work towards privileging and honouring Indigenous leadership at the forefront of One Health and Planetary Health project work.

[86] Redvers (2024).

[87] Id.

[88] Redvers et al. (2024), p. e0003634.

- Make space for listening and learning from Indigenous Peoples, and their Indigenous-led community examples of successful on the ground practice.
- Use a strengths-based lens over a deficit-based lens when working with Indigenous Peoples for research, policy, and practice.

6 Conclusion

The factors that contribute to public health risks are overlapping and multifaceted, existing across animal, human and environmental sectors. With the continuing threat of outbreaks from climate change and zoonotic origins, research that investigates the relationships between environmental and animal health is gaining momentum when it comes to their impacts on human health. If we are to accomplish the goal of sustained optimal human health, we must first start by placing a true priority on the wholistic health of the ecosystems and animals on which human health depends. Therefore, from our perspective, we must urgently work on addressing the root causes of climate change health disparities, and poor animal health with declining populations leading to zoonotic outbreaks. By tackling these risks to human health, we work towards a true Indigenous understanding of One Health and Planetary Health.

One Health and Planetary Health share many commonalities and favour wholistic approaches that recognise the importance of looking outside of the individual and acknowledging the shared environment to the benefit of health and wellbeing. These approaches are likely to continue to gain momentum with continuing health risks such as climate change, zoonotic disease, and antimicrobial resistance. Due to the intended wholistic nature of both One Health and Planetary Health, and the recognition of the need for approaching health challenges through desiloed and interdisciplinary approaches, they are likely to be an effective method to address health inequities within Indigenous populations if seeing Indigenous Peoples and their knowledges as co-partners. Substantial work is needed, however, to ensure Indigenous perspectives and opportunities are platformed within both the One Health and Planetary Health fields due to the current gaps in ensuring sustainable, effective, and appropriate interventions are implemented through policy, programs, and research fields.

References

Adisasmito WB, Almuhairi S, Behravesh CB, Bilivogui P, Bukachi SA, Casas N, Cediel Becerra N, Charron DF, Chaudhary A, Ciacci Zanella JR, Cunningham AA, Dar O, Debnath N, Dungu B, Farag E, Gao GF, Hayman DTS, Khaitsa M, Koopmans MPG, Machalaba C, Mackenzie JS, Markotter W, Mettenleiter TC, Morand S, Smolenskiy V, Zhou L, One Health High-Level Expert Panel (OHHLEP) (2022) One Health: A new definition for a sustainable and healthy future. *PLoS Pathogens 18*(6):e1010537. https://doi.org/10.1371/journal.ppat.1010537

Ali SH, Foster T, Hall NL (2018) The relationship between infectious diseases and housing maintenance in indigenous Australian households. Int J Environ Res Public Health 15(12):2827. https://doi.org/10.3390/ijerph15122827

American Public Health Association. (2024). Environmental Health. https://www.apha.org/topics-and-issues/environmental-health

Ancheta J, Fadaak R, Anholt RM, Julien D, Barkema HW, Leslie M (2021a) The origins and lineage of One Health, Part I. Can Vet J 62(8):883–885

Ancheta J, Fadaak R, Anholt RM, Julien D, Barkema HW, Leslie M (2021b) The origins and lineage of One Health, Part II. Can Vet J 62(10):1131–1133

Black Press Media Staff (2022) Research team is revitalizing Indigenous language with a first-of-its-kind blueprint. North Island Gazette https://www.northislandgazette.com/community/research-team-is-revitalizing-indigenous-language-with-a-first-of-its-kind-blueprint/

Boutsalis K (2020) The art of fire: reviving the Indigenous craft of cultural burning. The Narwhal. https://thenarwhal.ca/indigenous-cultural-burning/

Brand G, Wise S, Bedi G, Kickett R (2023) Embedding Indigenous knowledges and voices in planetary health education. Lancet Planetary Health 7(1):e97–e102. https://doi.org/10.1016/S2542-5196(22)00308-4

Burleigh A, McMahon S, Kiely S (2015) Owned dog and cat populations in remote Indigenous communities in the Northern Territory: a retrospective study. Aust Vet J *93*(5):145–150. https://doi.org/10.1111/avj.12321

Charron DF (2012) Ecohealth: origins and approach. In: Charron D (ed) Ecohealth research in practice. Insight and innovation in international development, vol 1. Springer, New York, NY. https://doi.org/10.1007/978-1-4614-0517-7_1

Christianson AC, Sutherland CR, Moola F, Gonzalez Bautista N, Young D, MacDonald H (2022) Centering indigenous voices: The role of fire in the Boreal Forest of North America. Curr For Rep 8:257–276. https://doi.org/10.1007/s40725-022-00168-9

Cobo JRM (1981) Study of the problem of discrimination against Indigenous populations: final report. United Nations Department of Economic and Social Affairs: Indigenous Peoples [Internet]. 1981 Jul [cited 2022 Jan 10]; Report No.: E/CN.4/Sub.2/476. https://www.un.org/development/desa/indigenouspeoples/publications/2014/09/martinez-cobo-study/

de Castañeda RR, Villers J, Guzmán CAF, Eslanloo T, de Paula N, Machalaba C, Zinsstag J, Utzinger J, Flahault A, Bolon I (2023) One Health and planetary health research: leveraging differences to grow together. Lancet Planet Health *7*(2):e109–e111. https://doi.org/10.1016/s2542-5196(23)00002-5

Faerron Guzman CA, Potter T, Aguirre AA, Astle B, Barros E, Bayles B et al (2021) The planetary health education framework. Planet Health Alliance. https://doi.org/10.5822/phef2021

Food and Agriculture Organization of the United Nations, United Nations Environment Programme, World Health Organization, World Organisation for Animal Health (2022) One Health joint plan of action (2022–2026): working together for the health of humans, animals, plants and the environment. Rome. https://doi.org/10.4060/cc2289en

Hill KX, Johson LJ, Blue MR, Probst J, Staecker M, Jennings LJ (2024) Rematriation and climate justice: intersections of Indigenous health and place. J Clim Change Health 18:100314. https://doi.org/10.1016/j.joclim.2024.100314

Hillier SA, Taleb A, Chaccour E, Aenishaenslin C (2021) Examining the concept of one health for indigenous communities: a systematic review. One Health 12:100248. https://doi.org/10.1016/j.onehlt.2021.100248

Hoffman KM, Christianson AC, Dickson-Hoyle S, Copes-Gerbitz K, Nikolakis W, Diabo DA, McLeod R, Michell HJ, Al Mamun A, Zahara A, Mauro N, Gilchrist J, Ross RM, Daniels LD (2022) The right to burn: barriers and opportunities for Indigenous-led fire stewardship in Canada. FACETS 7:464–481. https://doi.org/10.1139/facets-2021-0062

Lake FK, Wright V, Morgan P, McFadzen M, McWethy D, Stevens-Rumann C (2017) Returning fire to the land: celebrating traditional knowledge and fire. J For 115(5):343–353. https://doi.org/10.5849/jof.2016-043R2

Lock MJ, McMillan F, Bennett B, Martire JL, Warne D, Kidd J, Williams N, Roberts R, Worley P, Hutten-Czapski P (2022) Position statement: research and reconciliation with Indigenous Peoples in rural health journals. Rural Remote Health 22(1):7353. https://doi.org/10.22605/RRH7353

Lucero LJ, Cruz JG (2020) Reconceptualizing urbanism: Insights from Maya cosmology. Front Sustain Cities 2:1. https://doi.org/10.3389/frsc.2020.00001

Manitowabi S (2018) Medicine wheel teachings. In: Historical and Contemporary Realities: Movement Towards Reconciliation. Open Library. https://ecampusontario.pressbooks.pub/movementtowardsreconciliation/chapter/medicine-wheel-teachings/#:~:text=The%20Medicine%20Wheel%20reminds%20us%20that%20we%20need%20to%20balance,done%20to%20become%20re%2Dbalanced

Manuel G, Posluns M (2019) The fourth world: an Indian reality. University of Minnesota Press

Martens P (2023) Planetary health: the recipe for a sustainable future. Inaugural address. Maastricht University. https://pimmartens.com/wp-content/uploads/2023/05/inaugural-speech-pim-martens.pdf

Martens P (2024) Planetary health: the need for a paradigm shift. BioScience 74(3):128–129. https://doi.org/10.1093/biosci/biad107

McDonough C (2019) Will the river ever get a chance to speak: standing up for the legal rights of nature. Villanova Environ Law J 31:143. Available from: https://digitalcommons.law.villanova.edu/elj/vol31/iss1/4

Meisner J, McLeland-Wieser H, Traylor EE, Hermesh B, Berg T, Roess A, Van Patter L, Rosenthal A, Davidovitch N, Rabinowitz PM (2024) Relational One Health: A more-than-biomedical framework for more-than-human health, and lessons learned from Brazil, Ethiopia, and Israel. One Health. 2024 Jan 9;18:100676. https://doi.org/10.1016/j.onehlt.2024

Moreton-Robinson A (2015) The white possessive: property, power, and Indigenous sovereignty. University of Minnesota Press

Padhi A, Agarwal A (2024) To handle zoonoses better, Indigenous people must be included in policy making. Nature India. https://www.nature.com/articles/d44151-024-00127-2#:~:text=The%20active%20participation%20of%20Indigenous,for%20managing%20zoonotic%20diseases%20effectively

Parkes M, Waltner-Toews D, Horwitz P (2014) Ecohealth. In: Michalos AC (ed) Encyclopedia of quality of life and well-being research. Springer, Dordrecht. https://doi.org/10.1007/978-94-007-0753-5_4172

Planetary Health Alliance (2024) Planetary Health. https://www.planetaryhealthalliance.org/planetary-health

Redvers J (2020) "The land is a healer": perspectives on land-based healing from Indigenous practitioners in northern Canada. Int J Indigenous Health 15(1):90–107

Redvers N (2018) The value of global indigenous knowledge in planetary health. Challenges *9*(2):30. https://doi.org/10.3390/challe9020030

Redvers N (2021) Patient-planetary health co-benefit prescribing: emerging considerations for health policy and health professional practice. Front Public Health 9:678545. https://doi.org/10.3389/fpubh.2021.678545

Redvers N (2024) Indigenous planetary health and the bridging of Indigenous and conventional medicine systems. In: Routledge handbook of climate change and health system sustainability. Routledge. https://doi.org/10.4324/9781032701196-18

Redvers N, Poelina A, Schultz C, Kobei DM, Githaiga C, Perdrisat M, Prince D, Blondin B (2020) Indigenous natural and first law in planetary health. Challenges 11(2):29

Redvers N, Celidwen Y, Schultz C, Horn O, Githaiga C, Vera M, Perdrisat M, Mad Plume L, Kobei D, Kain MC, Poelina A, Rojas JN, Blondin B (2022) The determinants of planetary health: an Indigenous consensus perspective. Lancet Planet Health *6*(2):e156–e163. https://doi.org/10.1016/S2542-5196(21)00354-5

Redvers N, Menzel K, Ricker A, Lopez-Carmen VA (2023a) Blondin B. Expanding the scope of planetary health education: the International Decade of Indigenous Languages. Lancet Planet Health *7*(1):e4–e5. https://doi.org/10.1016/S2542-5196(22)00140-1

Redvers N, Aubrey P, Celidwen Y, Hill K (2023b) Indigenous peoples: traditional knowledges, climate change, and health. PLOS Global Public Health 3(10):e0002474. https://doi.org/10.1371/journal.pgph.0002474

Redvers N, Lokugamage AU, Barreto JPL, Bajracharya MB, Harris M (2024) Epistemicide, health systems, and planetary health: Re-centering Indigenous knowledge systems. PLOS Glob Public Health 4(8):e0003634. https://doi.org/10.1371/journal.pgph.0003634

Richardson K, Steffen W, Lucht W et al (2023) Earth beyond six of nine planetary boundaries. Sci Adv *9*(37):eadh2458. https://doi.org/10.1126/sciadv.adh2458

Rigney LI (1999) Internationalization of an indigenous anticolonial cultural critique of research methodologies: a guide to indigenist research methodology and its principles. Wicazo Sa Review 14(2):109–121. https://doi.org/10.2307/1409555

Riley T, Lovett R, Thandrayen J, Cumming B, Thurber KA (2020) Evaluating impacts of a one health approach to companion animal health and management in a remote aboriginal community in the northern territory, Australia. Animals (Basel) 10(10):1790. https://doi.org/10.3390/ani10101790

Riley T, Anderson NE, Lovett R, Meredith A, Cumming B, Thandrayen J (2021) One health in indigenous communities: a critical review of the evidence. Int J Environ Res Public Health 18(21):11303. https://doi.org/10.3390/ijerph182111303

Riley T, Anderson NE, Lovett R, Meredith A, Cumming B (2022) Zoonoses and the aboriginal and torres strait islander population: a one health scoping review. PLOS Glob Public Health 2(10):e0000921. https://doi.org/10.1371/journal.pgph.0000921

Roach P, McMillan F (2022) Reconciliation and Indigenous self-determination in health research: a call to action. PLOS Glob Public Health 2(9):e0000999. https://doi.org/10.1371/journal.pgph.0000999

Robinson JM, Gellie N, MacCarthy D, Mills JG, O'Donnell K, Redvers N (2021) Traditional ecological knowledge in restoration ecology: a call to listen deeply, to engage with, and respect Indigenous voices. Restorat Ecol *29*:e13381. https://doi.org/10.1111/rec.13381

Rousseau-Thomas J (2023) Indigenous Peoples have a rich tradition of thinking like good ancestors. Generation Squeeze. https://www.gensqueeze.ca/thinking_differently_using_the_seven_generations_principle

Sakakibara C (2009) 'No whale, no music': Iñupiaq drumming and global warming. Polar Record 45(4):289–303. https://doi.org/10.1017/S0032247408008164

Seven Generations Education Institute (2021) Seven Grandfather Teachings. https://www.7generations.org/seven-grandfather-teachings/

Sogorea Te' Land Trust (n.d.) What is Rematriation? https://sogoreate-landtrust.org/purpose-and-vision/

Stacey J, Verhaeghe C, Feltes E (2019) NAGWEDIẐK'AN GWANEŜ GANGU Ch'inidẑed GANEXWILAGH: The Fires Awakened Us: Tsilhqot'in Report on the 2017 Wildfires.. https://commons.allard.ubc.ca/fac_pubs/606/

Statistics Canada (2022) Population growth in Canada's rural areas, 2016 to 2021. https://www12.statcan.gc.ca/census-recensement/2021/as-sa/98-200-x/2021002/98-200-x2021002-eng.cfm

Talukder B, Ganguli N, Choi E, Tofighi M, Obrinski J (2024) Exploring the nexus: comparing and aligning Planetary Health, One Health, and EcoHealth. Glob Transit 6:66–75

Tauli-Corpuz V (2016) United Nations General Assembly: report of the special rapporteur of the Human Rights Council on the Rights of Indigenous Peoples. https://www.un.org/development/desa/indigenouspeoples/wp-content/uploads/sites/19/2019/01/N1624109.pdf

The United Nations Permanent Forum on Indigenous Issues (n.d.) Indigenous languages. https://www.un.org/development/desa/indigenouspeoples/wp-content/uploads/sites/19/2018/04/Indigenous-Languages.pdf

UN Environment Programme (2024) UNEP One Health. https://www.unep.org/topics/chemicals-and-pollution-action/pollution-and-health/unep-one-health

United Nations (2007) United Nations Declaration on the Rights of Indigenous Peoples. https://www.un.org/development/desa/indigenouspeoples/wp-content/uploads/sites/19/2018/11/UNDRIP_E_web.pdf

United Nations (2021) Secretary-General Calls Latest IPCC Climate Report 'Code Red for Humanity', Stressing 'Irrefutable' Evidence of Human Influence. https://press.un.org/en/2021/sgsm20847.doc.htm

University of Alberta (2024) First Nations, Métis, and Inuit- Indigenous Knowledges. https://guides.library.ualberta.ca/first-nations-metis-inuit/indigenous-knowledges

Waitangi Tribunal (2019) The stage 2 report on the national freshwater and geothermal resources claims. https://forms.justice.govt.nz/search/Documents/WT/wt_DOC_152208791/Freshwater%20W.pdf

Whitmee S, Haines A, Beyrer C, Boltz F, Capon AG, de Souza Dias BF, Ezeh A, Frumkin H, Gong P, Head P, Horton R, Mace GM, Marten R, Myers SS, Nishtar S, Osofsky SA, Pattanayak SK, Pongsiri MJ, Romanelli C, Soucat A, Vega J, Yach D (2015) Safeguarding human health in the Anthropocene epoch: report of The Rockefeller Foundation-Lancet Commission on planetary health. Lancet 386(10007):1973–2028. https://doi.org/10.1016/S0140-6736(15)60901-1

Whyte K (2017) Indigenous climate change studies: indigenizing futures, decolonizing the Anthropocene. English Language Notes 55(1):153–162

Yarra River Protection Ministerial Advisory Committee (2016) Protecting the Yarra river (Birrarung). Victoria Department of Environment, Land, Water and Planning. https://www.planning.vic.gov.au/__data/assets/pdf_file/0011/101711/DELWP0032.10_YarraRiverProtection_MACrecommendations_v14_weba.pdf

Yellow Cloud Q, Redvers N (2023) Honoring indigenous sacred places and spirit in environmental health. Environ Health Insights 17:11786302231157507. https://doi.org/10.1177/11786302231157507

The manufacturer's authorised representative in the EU is Springer Nature Customer Service Centre GmbH, Europaplatz 3, 69115 Heidelberg, Germany. If you have any concerns regarding our products, please contact ProductSafety@springernature.com

Printed and bound by CPI Group (UK) Ltd, Croydon, CR0 4YY
30/09/2026
02225592-0001